HARDWARE DESIGN AND SIMULATION IN VAL/VHDL

THE KLUWER INTERNATIONAL SERIES IN ENGINEERING AND COMPUTER SCIENCE

VLSI, COMPUTER ARCHITECTURE AND DIGITAL SIGNAL PROCESSING

Consulting Editor
Jonathan Allen

Other books in the series:

Relaxation Techniques for the Simulation of VLSI Circuits. J. White and A. Sangiovanni-Vincentelli. ISBN 0-89838-186-X.
VLSI CAD Tools and Applications. W. Fichtner and M. Morf, Editors. ISBN 0-89838-193-2.
A VLSI Architecture for Concurrent Data Structures. W.J. Dally. ISBN 0-89838-235-1.
Yield Simulation for Integrated Circuits. D.M.H. Walker. ISBN 0-89838-244-0.
VLSI Specification, Verification and Synthesis. G. Birtwistle and P.A. Subrahmanyam. ISBN 0-89838-246-7.
Fundamentals of Computer-Aided Circuit Simulation. W.J. McCalla. ISBN 0-89838-248-3.
Serial Data Computation. S.G. Smith, P.B. Denyer. ISBN 0-89838-253-X.
Phonologic Parsing in Speech Recognition. K.W. Church. ISBN 0-89838-250-5.
Simulated Annealing for VLSI Design. D.F. Wong, H.W. Leong, C.L. Liu. ISBN 0-89838-256-4.
Polycrystalline Silicon for Integrated Circuit Applications. T. Kamins. ISBN 0-89838-259-9.
FET Modeling for Circuit Simulation. D. Divekar. ISBN 0-89838-264-5.
VLSI Placement and Global Routing Using Simulated Annealing. C. Sechen. ISBN 0-89838-281-5.
Adaptive Filters and Equalizers. B. Mulgrew, C.F.N. Cowan. ISBN 0-89838-285-8.
Computer-Aided Design and VLSI Device Development, Second Edition. K.M. Cham, S-Y. Oh, J.L. Moll, K. Lee, P. Vande Voorde, D. Chin. ISBN: 0-89838-277-7.
Automatic Speech Recognition. K-F. Lee. ISBN 0-89838-296-3.
Speech Time-Frequency Representations. M.D. Riley. ISBN 0-89838-298-X.
A Systolic Array Optimizing Compiler. M.S. Lam. ISBN: 0-89838-300-5.
Algorithms and Techniques for VLSI Layout Synthesis. D. Hill, D. Shugard, J. Fishburn, K. Keutzer. ISBN: 0-89838-301-3.
Switch-Level Timing Simulation of MOS VLSI Circuits. V.B. Rao, D.V. Overhauser, T.N. Trick, I.N. Hajj. ISBN 0-89838-302-1.
VLSI for Artificial Intelligence. J.G. Delgado-Frias, W.R. Moore (Editors). ISBN 0-7923-9000-8.
Wafer Level Integrated Systems: Implementation Issues. S.K. Tewksbury. ISBN 0-7923-9006-7.
The Annealing Algorithm. R.H.J.M. Otten & L.P.P.P. van Ginneken. ISBN 0-7923-9022-9.
VHDL: Hardware Description and Design. R. Lipsett, C. Schaefer and C. Ussery. ISBN 0-7923-9030-X.
The VHDL Handbook. D. Coelho. ISBN 0-7923-9031-8.
Unified Methods for VLSI Simulation and Test Generation. K.T. Cheng and V.D. Agrawal. ISBN 0-7923-9025-3.
ASIC System Design with VHDL: A Paradigm. S.S. Leung and M.A. Shanblatt. ISBN 0-7923-9032-6.
BiCMOS Technology and Applications. A.R. Alvarez (Editor). ISBN 0-7923-9033-4.
Analog VLSI Implementation of Neural Systems. C. Mead and M. Ismail (Editors). ISBN 0-7923-9040-7.
The MIPS-X RISC Microprocessor. P. Chow. ISBN 0-7923-9045-8.
Nonlinear Digital Filters: Principles and Applications. I. Pitas and A.N. Venetsanopoulos. ISBN 0-7923-9049-0.
Algorithmic and Register-Transfer Level Synthesis: The System Architect's Workbench. D.E. Thomas, E.D. Lagnese, R.A. Walker, J.A. Nestor, J.V. Rajan, R.L. Blackburn. ISBN 0-7923-9053-9.
VLSI Design for Manufacturing: Yield Enhancement. S.W. Director, W. Maly, A.J. Strojwas. ISBN 0-7923-9053-7.
Testing and Reliable Design of CMOS Circuits. N.K. Jha, S. Kundu. ISBN 0-7923-9056-3.
Hierarchical Modeling for VLSI Circuit Testing. D. Bhattacharya, J.P. Hayes. ISBN 0-7923-9058-X.
Steady-State Methods for Simulating Analog and Microwave Circuits. K. Kundert, A. Sangiovanni-Vincentelli, J. White. ISBN 0-7923-9069-5.
Introduction to Analog VLSI Design Automation. M. Ismail, J. Franca. ISBN 0-7923-9071-7.
Principles of VLSI System Planning: A Framework for Conceptual Design. A.M. Dewey, S.W. Director. ISBN 0-7923-9102-0.
Mixed-Mode Simulation. R. Saleh, A.R. Newton. ISBN 0-7923-9107-1.
Automatic Programming Applied to VLSI CAD Software: A Case Study. D. Setliff, R.A. Rutenbar. ISBN 0-7923-9112-8.
Models for Large Integrated Circuits. P. Dewilde, Z.Q. Ning. ISBN 0-7923-9115-2.
Gallium Arsenide Digital Circuits. O. Wing. ISBN 0-7923-9081-4.

HARDWARE DESIGN AND SIMULATION IN VAL/VHDL

by

Larry M. Augustin
David C. Luckham
Benoit A. Gennart
Youm Huh*
Alec G. Stanculescu**
Stanford University

**Youm Huh is now with Samsung Electronics, Sunnyvale, CA, U.S.A.*
***Alec G. Stanculescu is now with Fintronics U.S.A., Inc.*

KLUWER ACADEMIC PUBLISHERS
Boston/Dordrecht/London

Distributors for North America:
Kluwer Academic Publishers
101 Philip Drive
Assinippi Park
Norwell, Massachusetts 02061 USA

Distributors for all other countries:
Kluwer Academic Publishers Group
Distribution Centre
Post Office Box 322
3300 AH Dordrecht, THE NETHERLANDS

Library of Congress Cataloging-in-Publication Data

Hardware design and simulation in VAL/VHDL / by Larry M. Augustin ...
[et al.].
p. cm. — (The Kluwer international series in engineering and computer science. VLSI, computer architecture, and digital signal processing)
Includes bibliography (p.) and index.
ISBN 0-7923-9087-3
1. VHDL (Computer hardware description language) I. Augustin, Larry M., 1962- . II. Series.
TK7885.7.H38 1991
621.39'2—dc20 90-5277
CIP

Printed on acid-free paper.

Printed in the United States of America

Contents

List of Figures

Preface

The VHSIC Hardware Description Language (VHDL) provides a standard machine processable notation for describing hardware. VHDL is the result of a collaborative effort between IBM, Intermetrics, and Texas Instruments, sponsored by the Very High Speed Integrated Circuits (VHSIC) program office of the Department of Defense, beginning in 1981. Today it is an IEEE standard (1076-1987), and several simulators and other automated support tools for it are available commercially.

By providing a standard notation for describing hardware, especially in the early stages of the hardware design process, VHDL is expected to reduce both the time lag and the cost involved in building new systems and upgrading existing ones.

VHDL is the result of an evolutionary approach to language development starting with high level hardware description languages existing in 1981. It has a decidedly programming language flavor, resulting both from the orientation of hardware languages of that time, and from a major requirement that VHDL use Ada constructs wherever appropriate.

During the 1980's there has been an increasing current of research into high level specification languages for systems, particularly in the software area, and new methods of utilizing specifications in systems development. This activity is worldwide and includes, for example, object-oriented design, various rigorous development methods, mathematical verification, and synthesis from high level specifications.

VAL (VHDL Annotation Language) is a simple further step in the evolution of hardware description languages in the direction of applying new methods that have developed since VHDL was designed.

VAL extends VHDL with a small set of new constructs. The purpose of VAL is to increase the capabilities of VHDL for (1) abstract specification, (2) hierarchical development of designs, and (3) validation. The new constructs in VAL are simple and easily understood by the VHDL user. VAL support tools are intended as additions to standard VHDL support environments so that the VHDL user can apply new methods within the VHDL context.

This book presents VAL assuming that the reader already has a working knowledge of VHDL. For example, familiarity with any one of the following books will suffice:

The VHDL Handbook. Coehlo, D., Kluwer Academic Publishers, 1989.

VHDL: Hardware Description and Design, Lipsett, R., Schaefer, C., and Ussery, C., Kluwer Academic Publishers, 1989.

VHDL Language Reference Manual. IEEE Standard 1076-1987, IEEE Press, 1989.

Chip-Level Modeling with VHDL. Armstrong, J., Prentice Hall, 1988.

This book is intended for two kinds of readers.

1. Those interested in methodology who want a simple overview of the kinds of new constructs in VAL and how they can be used in conjunction with VHDL,

2. Those interested in the principles of implementing validation tools based on VAL or similar kinds of specifications.

This book is structured into four parts.

Part I introduces VAL in a tutorial manner. It describes the concepts of VAL and explains their relationship to VHDL. This introduction emphasizes the principle of expanding the VHDL entity interface to include new VAL facilities for abstract specification. Modern techniques of specifying entities abstractly, and separately from any architectural implementation, are thereby introduced into the VHDL process. Other VAL concepts described in Part I include new features for specifying timing and for specifying hierarchical relationships between VAL abstract behaviors and VHDL architectural implementations. Part I introduces the concept of comparative simulation based on hierarchical specification and gives examples illustrating the use of VAL in the process of building VAL/VHDL descriptions.

Part II presents a selection of VAL/VHDL examples. The examples are graded in complexity from very simple one-level designs to examples with three levels of hierarchical development. A commentary accompanying each example highlights the various methods of using VAL that the example illustrates.

Part III serves as the VAL language reference manual. It describes the syntax and semantics of each VAL construct. This part provides a detailed insight into VAL and its relationship to VHDL.

Part IV describes an algorithm for translating VAL annotations into VHDL. This algorithm is presented informally with illustrations of each step. It is the basis for implementing VAL tools that automate comparative simulation of VAL/VHDL descriptions.

Parts I and II provide an overview for readers interested in concepts and methodology. Parts III and IV provide a handbook for those interested in design of VAL and implementation of tools supporting its use in validation of designs.

Finally, it is fair to say the VAL is not a complete or finished work. VAL represents a particular direction of evolution towards a new generation of hardware design languages. We believe that as language features for expressing abstraction and hierarchy develop, together with new methods of validation and synthesis, the programming language aspects of current hardware description languages will become unnecessary. Future design languages will eventually be simpler than current description languages. The ultimate purpose of this book is to encourage this kind of evolutionary process.

Acknowledgments

The authors are grateful for the support of the VHSIC program of the DOD under contract No. F33615-86-C-1137. The design of VAL and development of preliminary versions of the VAL support tools was undertaken under this contract. Bob Sahai and Jim Waters helped significantly in the implementation of the VAL tools and in developing examples.

HARDWARE DESIGN AND SIMULATION IN VAL/VHDL

Part I

A Tutorial Introduction to VAL

Chapter 1

Introduction

The VHSIC Hardware Description Language (VHDL) supports the design, description, and simulation of VHSIC components [23]. It provides a base language that can be used to describe hardware ranging from simple logic gates to complex digital systems. As an IEEE standard [25], VHDL will provide an important common base language for design tool development and design documentation.

VHSIC designs will incorporate anywhere from a few hundred to perhaps a million components. Managing this complexity requires a powerful support environment including a library manager, profiler, simulator, and other CAD tools. In addition, such environments must address the key problem of verifying the correctness of a design. If current practice continues, the VHDL designer will verify designs by using a simulator and manually comparing huge volumes of simulator output with an informal design specification. For large and complex designs, this approach is simply not practical.

CAD facilities for automating the comparison process, and integrating it with the simulation process, must be developed. To do this, the design specification itself must be expressed in a machine processable language such as an extension of VHDL. We shall refer to such specifications as *formal specifications*.

Generally, formal specifications omit (hide or abstract) some implementation details. Consequently, a specification will not completely constrain the behavior of the device, allowing the possibility of many different design implementations. Indeed, *separation* from implementation is a basic principle of formal specification. Separation of specification and implementation is supported by all modern programming languages.

By *verification*, we mean the process of determining that a design (an implementation, usually in the form of an architecture) of a device satisfies a separate formal specification of the device.

The ability to use separate formal specifications to verify design architectures places new requirements on existing CAD environments. Modern CAD environments for VHDL, and future design languages, must support more powerful verification techniques than are currently supported. Such techniques should provide precise information about what has actually been verified and where possible inconsistencies between a specification and an architecture may exist.

There are two commonly proposed approaches to developing new generation verification techniques:

- Comparative Simulation. A formal specification of the device is executed with an architecture and the two simulation results are automatically compared for consistency.

- Formal Consistency Proof. Mathematical proof techiques are used to prove that the design of the device satisfies a formal specification of the device's intended behavior.

In practice, neither approach should be expected to guarantee that a complex design is correct (that it satisfies a detailed formal specification). Comparative simulation would require simulating the design for all possible inputs, and formal proofs are too difficult to construct for all aspects of complicated designs. Rather, the objective of developing new CAD environments should be to support the use of both techniques in complimentary ways that increase the designer's confidence in the correctness of the design.

VHDL Annotation Language (VAL) [3, 4] provides an annotation facility that allows the VHDL designer to express simple kinds of specifications of device behavior during the design process. The goal of VAL is to make minimal extensions to VHDL to facilitate the use of modern abstract specification techniques, and the application of both comparative simulation and formal consistency proof to VHDL.

Annotation languages were first applied to the development and verification of software. In general, annotation languages for software permit information about various aspects of a program that is not normally part of the program itself to be expressed in a machine processable form [18]. They provide facilities for formally specifying both the intended behavior of the program and the properties of the problem domain upon which its correctness is based. They are intended to reduce programming errors by encouraging new methodologies of program construction and by providing a great deal of error checking at both compile time and run time. Readability of programs is also improved by enabling the programmer to express design decisions explicitly. Annotation languages also extend modern programming languages to support abstract specification of programs separately from any implementation.

These ideas still hold for an annotation language in the hardware description language domain. With this in mind, the design of VAL was undertaken with the following principle considerations:

1. VAL annotations should be simple enough that the extra time and effort required to annotate the design is more than paid for by the increased confidence in the correctness of the design and by the decreased debugging time.

2. VAL should not concern itself with the details of how a computation is performed, but rather with the specification of *what* computation is performed *when*.

3. VAL should be applicable to simulation time checking of VHDL, and should not affect the simulation result.

4. VAL annotations should be general enough to permit the use of formal abstract specifications during the design phase.

5. The designer should be free to annotate and specify as much or as little as desired.

6. It should be possible to apply formal proof techniques to prove that a VHDL design satisfies VAL annotations.

In VAL, new constructs in the entity interface provide a separate facility for abstract specification. A VAL entity interface is an abstract specification of *what* the entity does — what the functional and timing relationships between input and output data are, and what state changes occur. Specifications are separated from any architectural implementation.

The VHDL entity architecture provides an implementation of the behavior of the entity, that is, the architecture defines *how* the entity does what it is specified to do.

VAL extends VHDL with a small set of new constructs in three main categories:

1. constructs to specify (i) state, (ii) timing, and (iii) behavior in the entity interface,

2. constructs for defining concurrent processes that depend on one another,

3. annotations relating (or mapping between) abstract entity interface specifications and concrete architecture implementations.

By using VAL to specify an entity interface and VHDL to implement an architecture, the user of VAL/VHDL separates two concerns — *what* the entity does, and *how* the entity does it. The ability to specify behavior abstractly before committing to various implementation choices is thereby enhanced. Thus VAL encourages design processes that delay concrete choices until initial requirements are adequately represented in a top level behavioral specification.

VAL utilizes VHDL concepts wherever possible. This keeps new constructs to a minimum and allows new constructs to be easily understood given VHDL as a starting point. For example, the VHDL assertion facility already provides a simple kind of annotation capability. VAL assertions extend this feature to provide more freedom in specifying precisely when an assertion should hold within the VHDL timing model. Often the VHDL designer can express a given specification using either VAL or VHDL facilities. In such cases, VAL functions as a macro language extension of VHDL, providing more concise formulations.

In general, however, expressing abstract behavioral specifications at the entity interface level using VHDL features alone is often impossible, and usually awkward and complicated when it is possible. (We will analyze this problem with VHDL in more detail later.) VAL solves the problem by providing higher level constructs to express behavioral specifications more succinctly (categories 1 and 2 above), and to express relationships between behavioral specifications and architectural implementations (category 3 above).

VAL annotations have several possible applications, each of which may be supported by different environment tools. In this book we concentrate on applying VAL to (1) top-down design methodology, and to (2) automatic comparative simulation. Both applications are intended as approaches to managing development and verification of complex VHDL designs.

Other applications of VAL, such as formal consistency proof [6, 9, 7, 13, 1, 2], debugging [20, 21, 11], simulation optimization, and automatic synthesis of VHDL architectures from abstract behavioral specifications are current research topics.

In the area of verification, formal consistency proof holds much promise for future research. When practical, formal proof guarantees the correctness of a design without the need for exhaustive simulation. On the other hand, comparative simulation, while still limited by the problem of providing test vectors that adequately cover all aspects of device behavior, provides an easier and more currently practical approach to increasing confidence in the correctness of a design.

1.1 Comparative Simulation With VAL

A designer usually verifies a design using some form of simulation. This task often requires the designer to manually compare the simulation result with an informal design specification. Occasionally, the designer also has a high level behavioral description (written in, for example, C or Ada) whose output can be compared to the output of the simulator. The design is simulated using a set of test vectors, the behavioral model is run on the same test vectors, and the results are compared (Figure 1.1).

While this process of verification is adequate for simple designs, as designs become more complex it becomes less satisfactory. It is limited in the extent to which it allows the designer to debug a new design because it assumes a "black box" view of the design unit (or entity), in which the entity is accessible only through its ports.

Our model of design checking is based on utilizing a separate entity

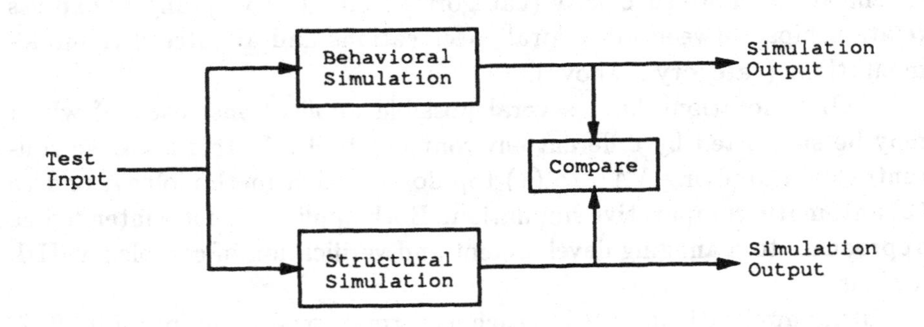

Figure 1.1: Typical Model of Design Checking

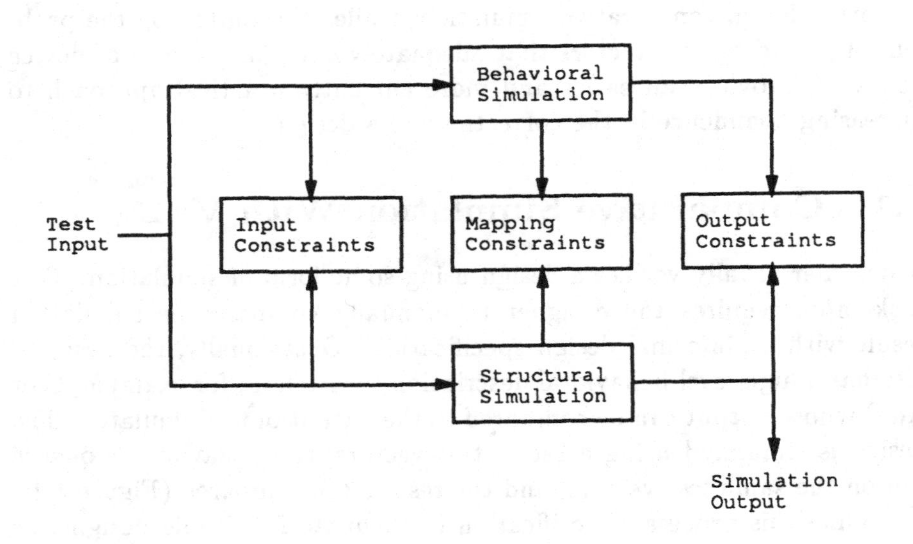

Figure 1.2: VAL Model of Design Checking

specification to generate constraints on the input of the entity, internal state, and output (Figure 1.2). Input constraints allow the simulator to check if an entity is being used correctly. For example, if the setup or hold time on a signal is not met, the entity can report an input constraint violation. This helps the designer to spot the source of timing errors as opposed to having to trace the source of the error back from the simulation result. Output constraints behave like the post simulation comparison previously described, with the addition that they may be executed dynamically, during the simulation. Mapping constraints allow an additional level of internal checking beyond the checking of ports. For example, if the behavioral description is a state machine, the states in the behavioral description must be somehow encoded within the structural model (i.e., distributed over the states of the lower level entities in the architecture.) Mapping constraints allow the designer to explicitly describe the encoding and allow the simulator to automatically check the internal state of the structure during simulation, rather than forcing the designer to deduce an incorrect state transition from the simulation result.

Support tools for comparative simulation with VAL can be engineered so that two simulations are run simultaneously; one being a VAL abstract entity interface specification, and the other being a VAL/VHDL entity architecture. The VAL mapping annotations relating the two permit automatic detailed comparison of the two simulations. Different architectures for a device can be automatically checked for conformance with an abstract entity specification. Furthermore, debuggers based on abstract specifications and mapping annotations can provide information about inconsistencies in terms of concepts used at the specification level.

Part IV of this book describes the implementation of one such tool, the VAL Transformer. The Transformer accepts as input a VAL/VHDL description and generates a self-checking VHDL program as output. It converts the VAL assertions in the VAL/VHDL description into VHDL statements. The VAL Transformer can be used as a front end to any VHDL simulator.

1.2 Why Extend VHDL?

The top-down approach to design development and verification that we shall describe does not depend on any particular hardware description language. However, its usefulness and effectiveness depends on the degree of support provided by the hardware description language for (1) separate abstract specification, and (2) definition of hierarchical relationships (mappings) between abstract specifications and implementations. Designers will not use a methodology if the language support is inadequate so that the methodology is complex and demanding.

VHDL contains many of the features needed to support our methodology. Thus, although VHDL was not designed to support abstract specification and comparative simulation, it provides a basis upon which to build a language extension that does.

Let us consider some of the kinds of extensions that are necessary.

Consider the problem of expressing the abstract (implementation independent) behavior of a device as part of the entity declaration in VHDL. What we would like to do is include in the interface specification (the entity declaration in VHDL) a set of specifications that define constraints on the input-output and timing behavior of the device. VHDL offers some support for this by allowing passive processes to appear in the entity declaration. However, the purpose of VHDL is to describe completed designs, not to support the design process. Consequently, VHDL does not include certain features that facilitate comparative simulation. One approach to simulation monitoring is the use of passive processes in the entity declaration. A passive process has all the features of the VHDL process statement with the exception that it cannot assign to a signal. The passive process was introduced as part of the refinement from VHDL 7.2 [24] to IEEE VHDL 1076/B to allow constraint checks applicable to all architecture bodies of an entity to appear in the entity declaration [26]. Passive processes may monitor the simulation, but may not change the state of the simulation. The user could attempt to use a passive process in the entity declaration to verify an implementation of a device. However, this would involve the user in constructing many additions to the original design whose sole function would be comparative checking, among which we mention the following:

1. Out mode ports may not be read within a device. This makes it

impossible to directly make assertions about the output generated by an architecture (the contribution of the entity to the value on the out port). The value placed on the out port may be accessed indirectly using a number of techniques. First, the architecture may be modified to place the value of the output port on a global signal at the same time the value is driven onto the output port. This approach is undesirable because it requires the designer to make modifications to the architecture which are not a part of the design of the device. Another alternative is to wrap the entity in a larger entity and capture the out mode signals on internal signals. This approach is used in the automated translation of VAL. It is, however, a complex problem, and prone to errors when generated by hand.

2. VHDL does not directly support the ability to make assertions about the stable value of signals. Because VHDL is discrete event simulation based, statements executed with "zero delay" actually execute in a unit delay time. Unit delay steps in the simulation often result as a consequence of the way the description is written, rather than because of some required behavior. Thus specifications typically refer only to the stable value of signals.

3. Assertions may not be guarded in VHDL. The guard associated with a VHDL block statement may guard only conditional signal assignment statements. Other concurrent statements (such as assertions) cannot be guarded. The result is that the user must construct a more complex form for the asserted expression that incorporates the guard. Requiring the user to code the guard conditions into the asserted condition is confusing and error-prone. Sections of texts on VHDL [5, 15] have been devoted to the problem of correctly writing such assertions.

4. The timing constraints in the specification of the device are often expressed in a form quite different than the VHDL description that results from those constraints. For example, in the TTL Data Book [27], edge triggered devices such as an LS74 are specified with a setup time measured *in the past* from the clock edge and a hold time measured *in the future* from the clock edge. Because

VHDL directly supports only references to the past value of signals (through the `'delayed` attribute), it is difficult to express this style of timing constraint in VHDL.

5. Because processes appearing in the entity declaration must be passive (they cannot modify signals) it is impossible to specify an abstract entity state in the entity interface and manipulate that state as part of the device behavior. If the behavior of the device can be captured in a single passive process, then state can be maintained using local variables. If, however, the device is modeled by multiple processes (which is necessary for describing some classes of behavior) then there is no way to communicate state information between passive processes.

While it is not impossible to construct a VHDL description such that an implementation may be automatically checked against a specification, doing so is awkward enough so as to preclude top-down design methodology and comparative simulation.

Thus VAL extends VHDL with the categories of annotations mentioned previously. In particular, in answer to each of the objections to VHDL above, the new VAL constructs support:

1. **Visibility over the contribution of out ports.** The designer can read the contribution of an entity to the value of its out mode ports.

2. **Flavors of assertions.** In addition to the VHDL assert, VAL supports other "flavors" of assertion statements that allow, for example, assertions to check only the stable value of signals.

3. **Hierarchical guarded statements.** Statements in VAL may be arbitrarily nested with guards. This simplifies the process of writing guarded assertions. The VAL preprocessor automatically translates the guards into the flattened form required by VHDL.

4. **Relative time.** VAL generalizes the idea of expressions involving time. The designer can reference the value of any expression at some time in the past or future relative to the current simulation time. This allows specifications to be written in a manner

corresponding directly to the style of specification currently used in manuals such as the TTL Data Book [27]. In addition, VAL includes constructs for determining the value of expressions over time intervals. This simplifies the specification of certain forms of timing behavior that depend on the value of signals over intervals of time.

5. **Abstract State model.** Ideally, it should be possible to model the state of a by an abstract data type. A high level specification might model a device state as a stack, set, or FIFO queue, for example. At a somewhat lower level, the specification of a state machine might use an abstract data type that was simply an enumeration of all the possible states plus a set of functions for transitioning between states. At a very detailed level, the abstract data type model for the state might consist of a set of state machines (components), together with other storage elements such as registers.

   As part of an entity specification, VAL allows the declaration of an instance of an abstract data type as the state model for a device. However, VAL abstract data types are restricted to the limited facilities available in VHDL for defining abstract data types. For example, VHDL packages cannot be generic, nor does VHDL support private types. As interfaces to other languages (such as Ada) from VHDL become available, VAL abstract data types can be programmed in languages which provide more powerful abstract type features. Until then, the ability to use abstract data types in VAL/VHDL will be limited.

1.3 Future Directions

Rather than viewing VAL as a completed language, we see VAL as a first step in developing languages to support the systems *design* process rather than just the *description* of completed designs. In the longer term, languages such as those called for in [8] will evolve to support the design process. In the near term, VAL may potentially evolve in a number of directions:

1. Language Independence – Although VAL was designed for annotating VHDL descriptions, many of the capabilities of VAL may be generalized for the specification of devices described in any language. With the exception of features such as visibility over out mode ports, flavors of assertions, etc., which are specific to VHDL, the remainder of VAL is simply a language for specifying behavior and timing constraints independent of any language for describing architecture. Other hardware description languages [14] could be used to describe the implementation of the device.

2. Formal Methods – Rules for proving that VHDL architectures satisfy VAL specifications can be defined. As a first example, timing constraints as expressed in VAL may be translated into Waveform Algebra [1]. Waveform Algebra is a formalism for representing timing and logic behavior. Because it has a basis in formal logic, it supports formal reasoning about timing constraints. VHDL statements also imply timing constraints that can be expressed in Waveform Algebra. Consequently, Waveform Algebra provides a common formalism in which the proofs that the timing of a VHDL architecture satisfies a VAL specification can be constructed.

3. Other Timing Models – VAL assumes VHDL global time semantics. Other timing models, appropriate to specifying distributed systems without a global clock, may be used in future versions of VAL.

1.4 Notation and Conventions

Throughout this book we will use the convention that keywords appear in bold, identifiers in roman, syntactic elements in italics, and comments in italics. Within BNF syntax diagrams, square brackets (`[ ]`) indicate optional parts, and curly braces ({ }) indicate a choice where a vertical bar (|) separates choices.

Chapter 2

An Overview of VAL

VAL extends VHDL by adding new constructs called *annotations*. This chapter presents a survey of the different kinds of VAL annotations.

A VAL/VHDL description is a VHDL description containing VAL annotations. Annotations are added as formal comments; lines beginning with --| are VAL annotations. Since comments in VHDL begin with two dashes, VHDL tools will interpret annotations as comments and ignore them. However, VAL tools will process them. This convention permits VAL/VHDL descriptions to be processed by standard VHDL tools as well as by VAL tools.

VAL annotations may appear in all three VHDL design units; entities, architectures, and configurations. We begin with a brief overview of the intended uses of annotations in each kind of design unit. The following sections will then explain the different kinds of annotations in detail.

Annotations in an entity declaration (*entity annotations*) can be used to define an abstract model of the state of the entity, and to specify relationships between inputs, outputs, and the entity state. In this way, a VAL/VHDL entity declaration specifies a behavior (or set of possible behaviors) to which all architectures for that entity must conform.

Thus, a VAL/VHDL entity declaration specifies a contract between a user of the entity and an implementor of the entity. A user of an entity (or an instance of it) can assume that the behavior of the entity satisfies the entity annotations. On the other hand, an implementor of an architecture for the entity is constrained so that the architecture will in fact satisfy those annotations.

The standard VHDL visibility rules regarding entity ports do not apply to annotations. These rules must be changed in order to permit entity annotations to be used effectively to define relationships between inputs and outputs. In VHDL, ports of mode out cannot be read within the entity and therefore assertions cannot normally be made about them. Unlike VHDL, VAL entity annotations have visibility over the contribution of the entity to each of its out ports. This allows the designer to specify the contribution of that entity to the output port. Constraints may also be defined on input ports. Static constraints on generic parameters are also declared as part of the entity annotations.

Annotations within an architecture body (*architecture annotations*) can be used to constrain the values of internal signals and ports of components. In addition, VAL annotations within the architecture have visibility over the abstract state of the entity (defined by the entity annotations) as well as the internal states of each component instantiated in the architecture. Annotations in an architecture relating to state are known as *mapping* annotations. In effect, mapping annotations describe the way in which the abstract state specified by the entity annotations is mapped into the set of states of the components of the architecture.

Annotations appearing in a configuration declaration (*configuration annotations*) allow the user to configure the VAL portion of the simulation. For example, the user may want to select only some of the entities in a large simulation for automatic checking. Also, the state model map, similar to a VHDL port or generic map, can be used to map an assumed state model for a component in an architecture into the actual state model of the actual component. This allows a designer to assume an abstract state model for a component that may not yet be available while designing an architecture and later provide a type conversion function to translate the assumed state model of the component to the state model of the actual component.

2.1 Entity Annotations

Figure 2.1 shows the layout of the VAL/VHDL entity declaration. The entity declaration consists of three parts; the entity header, the entity declarative part, and the entity statement part.

```
entity entity_name is
    generic(...);
    port(...);                  -- entity header
    --| state model is ...;

    type ...;
    subtype ...;
    --| assume ...;             -- entity declarative part
    --| type ...;
    --| subtype ...;
    --| macro ...;
begin
    assert ...;
    process ...;                -- entity statement part
    --| finally ...;
    --| when ...;
end entity_name;
```

The header contains the generic and port declarations. It may also contain a VAL state model declaration. The state model declaration declares the abstract data type to use in modeling the state of the entity.

The declarative part contains declarations (types, subtypes, etc.) local to the entity. VAL declarations (types, subtypes, macros, etc.) local to the VAL annotations associated with this entity appear here. Also, static boolean constraints (*assumptions*) appear in the declarative part. Assumptions typically express constraints on generic parameters of the entity, and so specify the set of contexts within which the entity may be used. Assumptions may be checked statically at compile time.

The statement part may contain VHDL passive processes. Annotations in the statement part are VAL *processes* that express constraints on the behavior of the entity. Processes in the statement part of the entity declaration may change the state of the entity and constrain the values appearing on the ports of the entity. This is in contrast to VHDL passive processes which may not modify any signals. VAL processes appearing here also have visibility over the value of out mode ports (actually, the contribution made by the entity to the value of the port).

Figure 2.2 shows a complete annotated VAL/VHDL entity declaration for a D flip-flop. Without trying to understand all the nuances of the description, familiarize yourself with the generic and port declarations. Also note that we have added annotations to the header (a state model declaration), the declarative part (an assumption), and to the statement part (some VAL processes). The D flip-flop will be used as an important example in the following sections. We will construct the annotations appearing in Figure 2.2 as we introduce the various language features of VAL.

2.1.1 Entity State Model

Because the future behavior of an entity may depend on its past behavior, VAL provides the entity state as a means of specifying history dependent behavior. The *state model* of an entity consists of a single object of any type allowed in VHDL. For example, records can be used to model complex states. The type of the state model is called the *state model type*. The state model type can be declared as an abstract data type in a separate VHDL package, along with the functions necessary to

```
-- DFlipFlop entity declaration
entity DFlipFlop is
  generic (SETUP, HOLD, DELAY : TIME);
  port (Clk  : in  bit;   -- Clock input
        D    : in  bit;   -- Data input
        Q    : out bit;   -- Output
        Qbar : out bit);

-- Assumptions about generics
--| assume (DELAY >= HOLD)
--|   report "Error in generic constant" ;

-- A single bit of memory
--| state model is bit := '0';
begin

-- State maintenance
--| when Clk'Changed('0') then
--|   when (D'Stable during [-SETUP, HOLD])
--|     then D -> State[DELAY] ;
--|   end when;
--| end when;

-- Check outputs
--| finally ((State = Q) and (not State = Qbar))
--|   report "Simulator error - D latch" ;
end DFlipFlop;
```

Figure 2.2: Annotated D Flip-Flop Entity Declaration

manipulate it, and imported into the VAL description. This allows the designer to import a generic package describing, for example, a Stack, Queue, Petri Net, or whatever abstract object most accurately models the state.

The state of the entity is visible to all of the entity annotations exactly as any VHDL declaration in the entity. Annotations in the entity declaration may read and assign to it. It is also visible to annotations in an architecture body of the entity but may only be read from within the architecture, exactly as if it were an out mode port of the entity. This allows the architecture to reference the state in making more detailed, implementation dependent annotations within the architecture. In addition, when an instance of the entity is used as a component in a parent architecture, the parent architecture has visibility over the state of the component as if the state were an additional out mode port of the component. This allows the parent architecture to make detailed implementation dependent assertions relating its state to the states of its components. Section 2.2 describes this use of the state in more detail.

The state model declaration has the following form:

```
--| state model is type_name;
```

Consider the specification of a D flip-flop. Figure 2.3 gives the VHDL entity declaration for a typical D flip-flop. The flip-flop has two inputs, `Clk` and `D`, and two outputs, `Q` and `Qbar`. It is generic in three timing parameters; the setup time (`SETUP`), the hold time (`HOLD`), and the delay through the flip-flop (`DELAY`).

The D flip-flop requires a single bit of memory to model its behavior. Its state model declaration is given by:

```
--| state model is bit := '0';
```

The keywords **state model is** indicate that the state is an object of the following type definition (in this example `bit`). An initial value for the state (in this example `'0'`) must be given. Later, as we develop the VAL constraints on the behavior of the D flip-flop, we will see how the state is used. Figure 2.4 gives the entity declaration including the state model declaration.

```
entity DFlipFlop is
    generic (SETUP, HOLD, DELAY : time);
    port (Clk  : in bit;
          D    : in bit;
          Q    : out bit;
          Qbar : out bit);
end DFlipFlop;
```

Figure 2.3: D Flip-Flop Entity Declaration

```
entity DFlipFlop is
    generic (SETUP, HOLD, DELAY : time);
    port (Clk  : in bit;
          D    : in bit;
          Q    : out bit;
          Qbar : out bit);
    --| state model is bit := '0';
end DFlipFlop;
```

Figure 2.4: D Flip-Flop Entity Declaration With State

2.1.2 Assumptions

Static constraints on generic parameters of an entity can be specified using the **assume** declaration. The **assume** declaration specifies conditions that should be observed by the user when making instantiations of the entity. Thus assumptions constrain the contexts or environments in which an entity may be placed. Any architecture of the entity may assume that these conditions are satisfied by the generic parameters. The **assume** declaration has a form similar to the VHDL assertion:

```
assume boolean_expression [else]
    [report expression]
    [severity expression];
```

The keyword **else** and the report and severity expressions are optional. The default severity level is **WARNING**. The boolean expression is assumed to be true, else an error message defined by the report clause is issued and the simulation may continue or be aborted depending on the severity clause.

Assumptions may appear in the declarative part of the entity declaration and may apply to generic parameters and constants. The condition assumed must be statically checkable at elaboration time.

Consider now the entity declaration of the D flip-flop we saw in Figure 2.4. The delay time through the device should be less than or equal to the hold time, otherwise the output of the device would have to change to a new value (after the delay time) before the stability requirement on the input (the hold time) was met. Clearly such a device is non-causal, and this will become more apparent when we specify the timing behavior of the flip-flop. We can make this constraint on the generic parameters explicit by stating it as an assumption:

```
--| assume (DELAY >= HOLD) else
--|   report "Error in generic constant";
```

If an instance of the D flip-flop is instantiated that violates this condition, a warning message (**"Error in generic constant"**) will be issued. Adding this assumption to the entity declaration yields Figure 2.5.

```
entity DFlipFlop is
    generic (SETUP, HOLD, DELAY : time);
    port (Clk  : in  bit;
          D    : in  bit;
          Q    : out bit;
          Qbar : out bit);
    --| state model is bit := '0';

    --| assume (DELAY >= HOLD) else
    --|     report "Error in generic constant";
end DFlipFlop;
```

Figure 2.5: D Flip-Flop Entity Declaration With Assumption

2.1.3 Statements and Processes

Annotations appearing in the entity statement part consist of a list of processes that execute in parallel. A process executes when it first becomes *active* or when it is active and there is an event on any of its signals. A process becomes active when its *guard* becomes true. A guard is a boolean expression which controls the activation of processes. There are two kinds of processes in VAL: assertions and drive. Statements control the creation of processes and their guards.

2.1.3.1 Assertions

An *assertion process* generates constraints on the simulation. Syntactically, it similar to the VHDL assertion:

```
{assert | finally | eventually | sometime } boolean_expr
    [else] [report expression]
           [severity expression];
```

The curly brackets ({ }) indicate that exactly one of the enclosing elements is permitted, and the vertical bar (|) separates choices. A VAL assertion process looks similar to a VHDL assertion process, but may in addition begin with the keywords **finally**, **eventually**, or **sometime**.

Consider the VHDL entity declaration for a two input AND gate shown in Figure 2.6. The identifiers `input_a` and `input_b` are input ports and `result` is the output port. Assertions in the form of VAL processes are added to define the behavior of this circuit. The behavior of the AND gate is specified by a single assert process that makes an assertion about the value carried on the output port. The VAL annotations describing the intended behavior of the entity appear following the VHDL keyword **begin**. The assert process checks a constraint (in this case `(input_a and input_b) = result`). It is similar to the assert statement in VHDL, but applies to a wider class of elements, such as output ports. If the constraint ever evaluates to false, the assert process performs the requested action. The **else** keyword emphasizes that the severity and report clauses are executed when the boolean expression is false. The **else** keyword is optional. The default severity level of VAL assertions is `WARNING`.

```
--  Annotated VHDL two input AND gate

entity TwoInputAND is
  port (input_a, input_b : in  bit;
        result           : out bit);
begin

--  VAL Annotations defining the AND gate's behavior

--| assert ((input_a and input_b) = result) else
--|    report "Error in TwoInputAND"
--|    severity FAILURE;

end TwoInputAND;
```

Figure 2.6: Annotated VHDL AND Gate Entity Declaration

VAL provides a family of assertion processes for generating constraints. The assert process is the strictest of these, requiring the constraint to be satisfied at every simulation cycle (i.e., at every delta). Desirable and adequate behaviors will often violate this constraint because a zero delay signal assignment in VHDL occurs after a delay of delta, limiting the usefulness of the VHDL assert statement for checking this kind of behavior. Other VAL assertion processes operate by generating constraints only at certain points during the simulation. For example, the finally assertion process allows the user to specify a constraint that must hold only at the last delta in a simulation time point. The constraint generated by the annotation in Figure 2.6 will report an error for a single delta whenever a change in **input_a** or **input_b** causes a change in **result** because the VHDL simulation of any architecture for this entity does not effect the change until the next delta. Replacing **assert** by **finally** checks only after the assignment has been completed, reflecting more closely the intentions of the designer. In the remainder of this chapter we will use the finally assertion exclusively because we are only interested in the final stable value of signals, and not in the intermediate steps in the simulation cycle. More details about the various flavors of assertions supported in VAL and their relationship to the VHDL simulation cycle can be found in Section 3.2.2.

2.1.3.2 Drive

In the behavioral description, the keyword **state** refers to the state model. The VAL drive process changes the state. It defines a process that, when executed, assigns the value of an expression to an object. Whenever the value of the expression changes, the value of the object also changes. Drive can only be applied to a component of the state. When the state of a device is not driven, it retains its last value. Syntactically, a drive process is expressed as:

```
expression -> state_object;
```

or,

```
state_object <- expression;
```

The first alternative reads from left to right as "expression *drives* state_object," and the second appears more like a conventional signal assignment statement.

2.1.3.3 Guards

Unlike VHDL, VAL provides the ability to hierarchically nest assertions within guards. The keyword **when** identifies two lists of processes, corresponding to a **then** part and an **else** part, and a boolean guard expression. Syntactically it is similar to an if statement:

```
when boolean_expression
[then statement_list]
[elsewhen statement_list]
[elsewhen statement_list]
        :
[else statement_list]
end when;
```

If the boolean expression evaluates to true, the processes following the **then** clause are activated, otherwise those following the **else** clause are activated. Note that any change in the value of the guard results immediately in the activation of one branch and the deactivation of the other branch. Each guard can be viewed as a node in a binary tree, the two branches being its **then** and **else** parts. A process at any level in the tree is active if and only if all of the guards leading to that point in the tree are true.

Nested guards, **else** clauses, and **elsewhen** clauses can all be flattened into a set of simple (no else or elsewhen parts) guards. For example:

```
when e1 then
    s1;
elsewhen e2 then
    s2;
else
```

```
    s3;
end when;
```

is equivalent to:

```
when e1 then s1; end when;
when not e1 and e2 then s2; end when;
when not e1 and not e2 then s3; end when;
```

Nested guards behave as if they were flattened using conjunctions:

```
when e1 then
    when e2 then
        s2;
    end when;
end when;
```

is equivalent to:

```
when e1 and e2 then
    s1;
end when;
```

Thus each process can be viewed as having a single guard controlling it. An unguarded process is equivalent to a process guarded by the constant `TRUE`.

Consider for the moment how the D flip-flop behavior might be specified if there were no timing constraints. Using a guarded drive process we can construct a VAL process that changes the state of the D flip-flop to the value on the `D` input port when the `Clk` input changes from `'1'` to `'0'` (a falling edge trigger).

```
--| when (Clk = '0') and (Clk'event) then
--|     D -> state;
--| end when;
```

Whenever the clock input is `'0'`, and an event has occurred on the clock input, the internal state of the D flip-flop follows the value on the data input. Otherwise, the state maintains its last driven value.

Because the "edge-triggered" behavior of the flip-flop is so common in the description of hardware devices, VAL contains an additional signal attribute **'changed()** to make expressing this behavior simpler. The attribute **S'changed(expression)** is **TRUE** if the value on signal **S** has changed to the value of the expression at the current time instant. It is equivalent to **(S = expression) and S'event**. The process for the D flip-flop can be rewritten using the **'changed** attribute:

```
--| when Clk'changed('0') then
--|     D -> state;
--| end when;
```

Of course, this assertion ignores timing behavior. Later, when we have explained annotation of timing behavior in VAL, we will see how this assertion is modified to handle timing.

2.1.3.4 Select

Select offers a more convenient syntax for activating a process based on the value of an expression of a discrete type.

```
select expression is
     in choice {| choice} => statement_list
     in choice {| choice} => statement_list
                    :

    [in others => statement_list]
end select;
```

A select statement is semantically equivalent to a set of mutually exclusive when statements. The expression must be a discrete type, and the type of each of the choices must match the type of the expression. Each set of choices must be mutually exclusive.

2.1.3.5 Generate

The generate statement creates a set of processes differing only in a single parameter. The syntax of a VAL generate statement is the same as that of the VHDL generate statement:

```
for identifier in range generate
     statement_list
end generate;
```

Semantically, the result of a generate statement is equivalent to textually creating a copy of each of the statements in the statement list with the generate parameter replace by its value for that process.

2.1.3.6 Macro Call

A macro is used in abstracting a common behavior that appears at multiple places in annotations. A macro call statement instantiates a previously declared macro. Macros are declared in the entity declarative part or in packages. (See Section 14.4 of the VAL Language Reference Manual.) The instance of a macro is expanded textually with the corresponding macro body during the VAL preprocessing.

```
macro_name( parameter_list );
```

When a macro is declared with parameters, a macro call statement has to provide arguments that match the macro parameters. Arguments may be designated by position or by name. Arguments for all parameters must be specified; default arguments for parameters are not allowed. Parameters are substituted for corresponding arguments during macro expansion.

It is prohibited to call a macro recursively.

2.1.3.7 Function Call

VHDL functions may be called from VAL annotations. This means that VAL has the full power of the sequential part of VHDL available for performing standard programming tasks. VHDL functions called from VAL use the same syntax as VHDL.

2.1.4 Timing Behavior

In the following discussion, the particular unit of time has been neglected since the units are irrelevant to the discussion and serve only to clutter the examples. VAL requires (as does VHDL) that all references to time be of physical type **TIME**.

A VAL/VHDL signal is a sequence or *stream* of values ordered by time. Conceptually, an entity specification defines a mapping between a set of input streams (values on its **in** ports) and a set of output streams (values on its **out** ports). Each expression in VAL defines a stream of values; expressions refer to *relative* time points on their streams of values. The relative time *-t* refers to a point in time on the stream that occurs *t* units before (or in the past) relative to the current time. Similarly, the relative time *t* refers to a point in time on the stream that occurs *t* units after (or in the future) relative to the current time.

2.1.4.1 Timed Expressions

Timed expressions in VAL are simple functions of time that allow the designer to describe relative timing relationships. The timing operator **[ ]** can be applied to any object or expression to refer to its value at any relative point in time. Thus **signal_a[-5]** refers to the value of **signal_a** five time units ago, and **signal_a[5]** refers to the value of a five units in the future. The expression **signal_a** refers to the current value of **signal_a** and is equivalent to **signal_a[0]**.

Assertion and drive processes define relationships between streams described with timed expressions. Consider the following drive process:

```
signal_a[-3] -> signal_b;
```

This is similar to the delayed assignment of VHDL (**signal_b <= signal_a after 3;**) except that the semantics of the VAL process are anticipatory [17, 2] rather than preemptive (as in VHDL). (See Section 3.2 for a discussion of anticipatory delay verses other delay models.) In VAL this process describes a relationship between the stream associated with **signal_a** and the stream associated with **signal_b**. An equivalent process would be:

```
signal_a[-1] -> signal_b[2];
```

This describes the same relationship between the streams; the value on **signal_a** has the same shape as **signal_b**, but is shifted 3 time units into the past.

VAL cannot change the past value of an object. Therefore, non-causal processes such as,

```
signal_a[5] -> signal_b[-1];
```

(the present value of **signal_b** is determined by the value of **signal_a** 6 time units in the future) which have no physical meaning are not allowed.

2.1.4.2 Time Intervals

The timing qualifier **during** can be applied to check the value of a boolean expression over a time interval. The expression

```
(signal_a = 1) during[-3,2]
```

is true if **signal_a = 1** from 3 units ago to 2 units in the future; otherwise it is false. Since timing qualifiers commonly refer to a range over the most recent interval, the expression

```
(signal_a = 1) during 5
```

is a shorter notation for

```
(signal_a = 1) during[-5,0]
```

2.1.4.3 Choosing a Reference Point

All expressions in VAL refer to some time point – expressions that are not timed refer to the current simulation time. The *reference point* of an entity specification is that time value chosen as the current time. For example, to express "D must be stable for X time units" with the reference point $t = 0$, we write

```
D'stable during[-X, 0]
```

If the reference point is $t = 1$ we write

```
D'stable during[-X + 1, 1]
```

The choice of reference point important in simplifying the timing description in a specification and determining causality of a specification.

As an example of timing behavior, we return again to the D flip-flop example. Informally, a description of this device will typically include a propagation delay as well as setup and hold time requirements on the data. Assume that the state of the D flip-flop is only changed if the data remains stable during a time **SETUP** before the falling edge of the clock and a time **HOLD** after the falling edge. Also, assume that the new value of the data takes **DELAY** time units to appear on the output after the state has changed.

Figure 2.7 shows a typical timing diagram for the D flip-flop. To express these requirements, we choose $t =$ **clk'changed('0')**, the time at which the clock edge falls, as our reference point. The VAL specification (Figure 2.2) follows almost exactly from the above informal description.

There are two concurrent processes at the top level in the VAL specification:

1. Maintain state:

```
--| when Clk'changed('0') then
--|   when (D'stable during [-SETUP, HOLD])
--|     then D -> State[DELAY];
--|   end when;
--| end when;
```

2. Check outputs:

```
--| finally ((State = Q) and (not State = Qbar))
--|   report "Simulator error - D latch";
```

Both of them execute concurrently. Process (1) defines a behavior dependent on the clock input value. When the clock changes from some other value to `'0'` the guard becomes true, and the nested guard is activated. If the D input has been stable in the interval (`-SETUP`, `HOLD`) about the current time point, then the nested guard expression is true, and the drive process is activated. The drive process changes the value of the state at a time `DELAY` in the future relative to the current value of the D input.

Process (2) is a finally assertion about the abstract state value and the output ports `Q` and `Qbar`. The assertion, being unguarded, is always active. Consequently, process (2) constrains the state and output port `Q` to be equal at the end of every time point in the simulation for any architecture body associated with this entity. The assertion actually constrains only the value placed on the output port by an architecture associated with the D flip-flop, since the port may be connected to a bus whose value is some function of all the ports connected to it. The VAL specification says nothing about the value carried by such a bus.

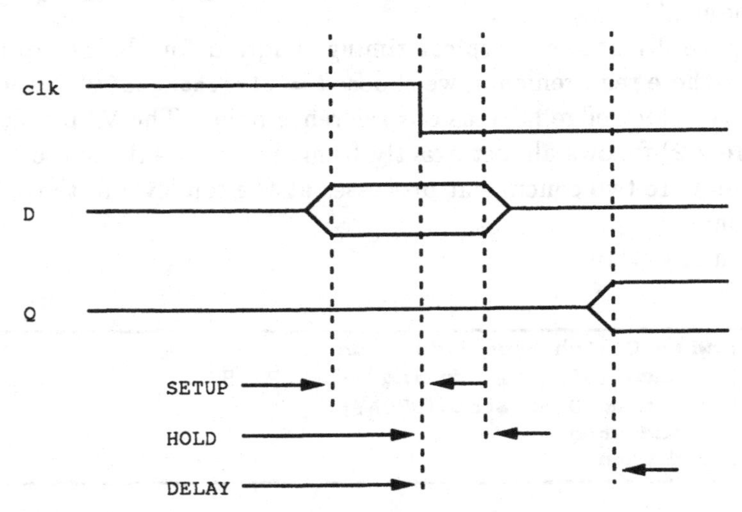

Figure 2.7: D Flip-Flop Timing Diagram

When `Clk'changed('0')` becomes true, the guarded process checking the setup and hold time of the data becomes active. Note that the expression **during**[-SETUP,HOLD] checks the interval SETUP time units in the past and HOLD time units in the future. If the data remains stable over this interval, the internal state of the D flip-flop is modified after a time DELAY. The assertion processes constrain the ports of the VHDL body to match the state bit, and its negation, at all times.

The constraint DELAY >= HOLD is worth exploring further. Consider any time point t at which Clk'**changed**('0') is true. The inner nested process is activated. Taking some liberties with the VAL notation, the signal D is then checked over the interval [t - SETUP, t + HOLD]. If it is stable over the entire interval (i.e., if D'**stable** **during**[t - SETUP, t + HOLD] is true) then the drive process (D -> **State**[DELAY]) is activated and the value of the signal **State** at the time point t + DELAY is assigned the value of D. The description of the behavior of the D flip-flop is (conceptually) executed for every time point t in the simulation. The assignment to state cannot happen if the time point t + DELAY has already passed; i.e., t + HOLD > t + DELAY.

Conceptually, this implies that the output can never take on a new value before that new value is latched into the internal state. If DELAY < HOLD were true, then the output could change after DELAY time units, but the hold constraint might not yet be met, in which case the output value should never have changed. In other words, if DELAY < HOLD then the behavior is said to be non-causal. This is more obvious if the VAL description in Figure 2.2 is rewritten such that the reference point is the point at which the state is assigned a new value. The relevant lines in Figure 2.2 become,

```
when D'stable during[-SETUP-DELAY,HOLD-DELAY] then
  D[-DELAY] -> state[0];
end when;
```

If HOLD - DELAY > 0, then the assignment to the new value of state depends on an event that hasn't happened yet – the stability of the input during the hold time.

2.1.4.4 Loose Timing Constraints

It is not always the case that we can assign an exact time to the delay in a specification. Typically, a timing constraint will indicate a range of allowed values within which the constraint must hold. The timing qualifier **within** supports this style of constraint in VAL by checking if a boolean expression is true somewhere in an interval. The expression

```
signal_a within [-3, 2]
```

is true if **signal_a** is true somewhere in the interval from −3 units in the past to 2 units in the future.

Consider now the problem of "loosening" the specification given in Figure 2.2. The last assertion requires that the outputs always change at exactly the same time as the state. Certainly it would be almost impossible to design a D flip-flop meeting this specification. Suppose that we wish to relax the specification by allowing the outputs to change to the new value slightly early (e.g., 2 fs) or slightly late (e.g., 5 fs). The relaxed specification can be written in VAL as:

```
finally ((state = Q) and (not state = Qbar))
    within [-2 fs, 5 fs];
```

The assertion holds at time t if the expression `(state = Q) and (not state = Qbar)` is true somewhere in the interval `[`t `+ -2 fs,` t `+ 5 fs]`. Note that to ensure causality, we must now have `HOLD - DELAY - 2 > 0`, otherwise we allow the output to change before the hold time constraint is met.

The **within** timing qualifier can be expressed in terms of **during**:

```
signal_a within[T1,T2] =
    not ((not signal_a) during [T1, T2])
```

The two qualifiers are thus duals:

```
not (signal_a within[T1,T2]) =
    (not signal_a) during [T1, T2]
```

The **within** timing qualifier provides a succinct notation for expressing loose timing constraints.

2.2 Architecture Annotations

Any of the VAL processes, with the exception of drive, can appear in an architecture body. *Architecture annotations* specify implementation details and allow more detailed consistency checking between the entity annotations (the behavioral specification of the entity) and the VHDL architecture body (implementation).

To express detailed relationships between entity specifications and architectures, architecture annotations must have visibility over signals not normally visible to VHDL assertions appearing at the same point. In particular, they must have visibility over the abstract state of the entity defined in the entity declaration, and the abstract state of any components.

VAL makes this possible by providing visibility over the abstract state as if it were an out mode port. That is, it may be read but not assigned within an architecture. Thus architecture annotations have visibility over all VHDL signals and ports normally visible at the point at which the annotation appears, the state model of the entity, and the state models of all entities instantiated as components.

Figure 2.8 shows the relationship between entity and architecture annotations. The architecture in the figure contains two components connected by several internal signals (wires). Each component consists of an interface (entity declaration), entity annotations (specification), and an architecture body (implementation). The entity annotations in a component have visibility over the signals connected to the architecture of the component. That is, they have visibility over the ports of the component. In addition, the entity annotations make the state model visible to both the architecture of the component, and the parent architecture using an instance of the component. In Figure 2.8, the solid lines indicate the signals present in the VHDL description, and the dotted lines indicate the additional visibility available to the VAL annotations.

Note that the visibility of state applies hierarchically to nested components in a design; the architectures of the components may contain annotations and other components. Architecture annotations, however, may not modify the state model. Only the entity annotations may do so.

Because architecture annotations have visibility across two levels in

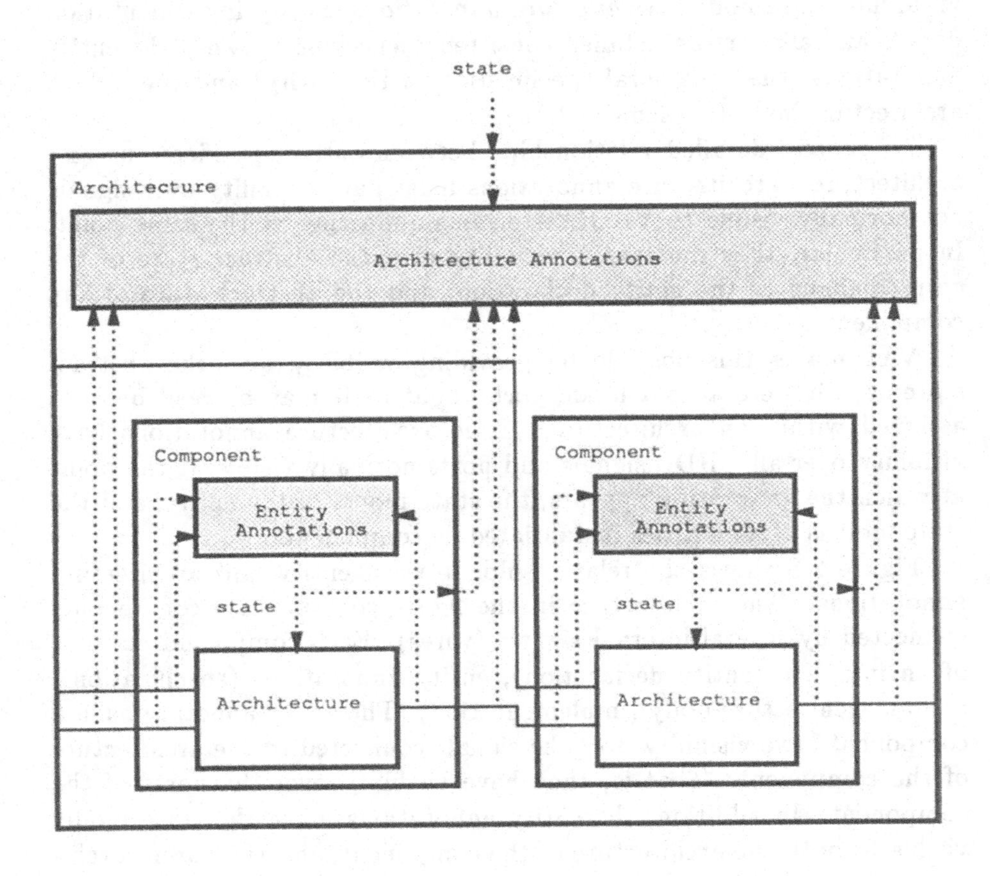

Figure 2.8: Hierarchical Annotations

the hierarchy, they may express constraints relating the state model of a device to the state model(s) of its component(s). Annotations that relate state models are called *mapping* annotations because they provide a simple mechanism for describing how the state of the device is mapped onto the state of its components.

```
entity TwoBitCounter is
  port (Clk           : in bit;
        reset         : in bit;
        Bit0, Bit1    : out bit);

--| state model is integer := 0;

begin

--| when reset then
--|   state <- 0;
--| elsewhen Clk'changed('0') then
--|   state <- (state + 1) mod 4;
--| end when;
--| select state is
--|   in 0 => finally(Bit0 = '0' and Bit1 = '0')
--|          report "Counter - Output error"
--|          severity warning;
--|   in 1 => finally(Bit0 = '1' and Bit1 = '0')
--|          report "Counter - Output error"
--|          severity warning;
--|   in 2 => finally(Bit0 = '0' and Bit1 = '1')
--|          report "Counter - Output error"
--|          severity warning;
--|   in 3 => finally(Bit0 = '1' and Bit1 = '1')
--|          report "Counter - Output error"
--|          severity warning;
--| end select;

end TwoBitCounter;
```

Figure 2.9: Two-bit Counter Entity Declaration

The description of the two-bit modulo four counter in Figure 2.9 and Figure 2.10 together show how mapping annotations may be used to check the internal state of an entity. The **reset** signal sets the state of

```
architecture SIMPLE of TwoBitCounter is

    signal Q1, Q2, Q1bar, Q2bar : bit;
    signal D1, D2 : bit;

    component DFlipFlop
      port(Clk, D : in bit;
           Q, Qbar : out bit;
           Reset : in bit);
      -- local state model declaration
      --|  state model is bit;
    end component;

  begin
    DFL1 : DFlipFlop
      port map ( Clk, D1, Q1, Q1bar, Reset);
    DFL2 : DflipFlop
      port map ( Clk, D2, Q2, Q2bar, Reset);
    D2 <= (Q1 and Q2bar) or (Q1bar and Q2);
    D1 <= Q1bar;
    Bit0 <= Q1;
    Bit1 <= Q2;

-- mapping annotations relate the state of the counter
-- to the states of the components

--| select state is
--|   0 => finally ( DFL2.state = '0' and DFL1.state = '0')
--|     report "Counter and flip-flop state mismatch"
--|     severity warning;
--|   1 => finally ( DFL2.state = '0' and DFL1.state = '1')
--|     report "Counter and flip-flop state mismatch"
--|     severity warning;
--|   2 => finally ( DFL2.state = '1' and DFL1.state = '0')
--|     report "Counter and flip-flop state mismatch"
--|     severity warning;
--|   3 => finally ( DFL2.state = '1' and DFL1.state = '1')
--|     report "Counter and flip-flop state mismatch"
--|     severity warning;
--| end select;

end SIMPLE;
```

Figure 2.10: Two-bit Counter Architecture

the counter. Whenever a transition from '1' to '0' on the clock (Clk) occurs, the counter counts up one. Bit0 represents the least significant bit of the counter and Bit1 the MSB. The VAL state model is an integer and assert processes generate constraints on the output ports based on the VAL state.

The architecture SIMPLE of the counter contains two D-type flip-flops. Each flip-flop is similar to the ones described previously with the exception of a reset signal and the omission of timing information (to keep the example short). Each flip-flop has a state model consisting of a single bit. The states of the flip-flops (DFL1.state and DFL2.state) are related to the state of the counter (state) by mapping annotations.

2.3 Configuration Annotations

Configuration annotations serve two purposes. First, they provide a local state model mapping declaration to map the local state model defined in a component declaration to the actual state model defined by the entity annotations of the components. The state model mapping declaration indicates the function to use in mapping between the state model of the actual entity and the state model of the component instance. It appears within a configuration specification at the same point as other binding indications.

```
use work.all;
configuration A of TwoBitCounter is
  --| valentity;
  --| valarchitecture;
  for SIMPLE
    for all: DFlipFlop use
      entity DFlipFlop(behavior_body);
      --| valentity;
    end for;
  end for;
end A;
```

Figure 2.11: Two-bit Counter Configuration

Second, they provide configuration information so that VAL generated architectures may be automatically substituted for original component architectures for checking. The user may not want to use a VAL annotated entity in place of the original VHDL entity for all components in a simulation, particularly if the component is a library unit for which no annotated description exists. The **valentity** construct allows the user to select the components of an architecture to be monitored. The VAL Transformer will only generate code to monitor components marked with **valentity**. Part IV of this book on the VAL Transformer explains how components are monitored.

In Figure 2.11 we have the following configuration declaration for the **DFlipFlop**:

```
for all: DFlipFlop use
  entity DFlipFlop(behavior_body);
  --| valentity;
end for;
```

This declaration tells the VHDL simulator to use the entity named **DFLipFlop** and the architecture of that entity named **behavior_body** for every instance of a DFlipFlop component. The **valentity** flag tells the VAL Transformer to enable the annotations appearing in the **DFlipFlop** entity. If there are any annotations in architecture **behavior_body** of entity **DFlipFlop**, they will be ignored. If the **valarchitecture** flag were present for the **DFlipFlop**, then the annotations appearing in the architecture **behavior_body** would be enabled, and they would be used to perform runtime checking.

For the **TwoBitCounter**, Figure 2.11 specifies both the **valentity** and **valarchitecture** flags. This tells the VAL Transformer to translate the annotations in both the entity declaration of **TwoBitCounter** and the architecture **A** of entity **TwoBitCounter**.

Chapter 3

Timing Models

One of the major differences separating programming languages and hardware description languages is the need to specify timing. The temporal behavior of a program is usually not a fundamental part of an algorithm, whereas the temporal behavior of hardware is often a critical part of the requirements, and must be expressed even in an initial specification. The need to express timing in hardware languages is complicated by the fact that there is no standard set of language constructs for temporal specifications. The area of timing, semantics of timing constructs, and temporal logics is still very much a research area.

In this chapter, we first review the timing model of VHDL [25]. The VHDL model is quite complex. VHDL employs preemptive timing statements – the effect of a timing statement is not known when it is executed because it may be preempted by another statement executed later. Two different preemptive delay models are supported: transport and inertial.

VAL is based on a simpler (and more abstract) timing model that hides some of the details of the VHDL model. After describing the semantics of the two VHDL models, we present the *anticipatory* timing model of VAL [2].

A statement is said to be *anticipatory* if its effect takes place *after* it is executed, but once the statement is executed its effect must take place. The statement "anticipates" the future. Thus, in VAL, the effect of a statement is never preempted (or "undone") once the statement is executed. Anticipatory semantics of VAL timing constructs permits the manipulation of VAL timing specifications using formal techniques.

This enables a designer to reason about VAL timing specifications, and prove that they express desired properties, before using them to check a VHDL architecture. Similar techniques for reasoning about VHDL timing behavior are not known at present.

The VAL support tools check simulations of architectures in VHDL (i.e., architectures that use VHDL preemptive timing constructs) for consistency with VAL specifications that include timing (using VAL anticipatory timing constructs).

3.1 The VHDL Timing Model

Despite its name, VHDL is not specifically a *hardware* description language, but is rather a *discrete event simulation* description language. The only description of the semantics of its timing operations is given in terms of a discrete event simulator. An event corresponds to a change in value of a signal, and may project other events in the future. That is, given a current set of values for its signals, a VHDL description predicts the expected output at future points in time. As time advances, projected outputs become actual outputs, at which point they are propagated along data paths to become inputs to components, where they may possibly project more future outputs [22].

The event driven semantics of VHDL are based on the assumption that all signals propagate in well-defined directions and that signal propagation always involves some delay; there is no instantaneous change of value. Descriptions (or portions thereof) that do not require complete timing details may use a unit delay model of computation. Descriptions which specify delays use either inertial or transport delay mechanisms.

3.1.1 Unit Delay

VHDL has a two-level hierarchy of time scales. The *macro* scale measures real time, and is typically the time unit of concern to the user. The smallest unit of macro time measurable in VHDL is 1 femtosecond (fs). Macro units are referred to as *time points.* The micro scale measures "unit delay" time. Any number of micro-units (each micro-unit is called a *delta*) may exist between any two time points. In VHDL, each delta corresponds to a computation step in the simulation. Hence, in VHDL one delta is equivalent to one *simulation cycle.* In this book, we often use the two terms interchangeably, although we try to use the term delta in the context of an infinitesimal delay, and the term simulation cycle in the context of a discrete event simulation. The set of all simulation cycles at a given simulated time corresponds to a time point. Figure 3.1 shows the relationship between the scales. Each time point is separated by a (possibly infinite) number of deltas.

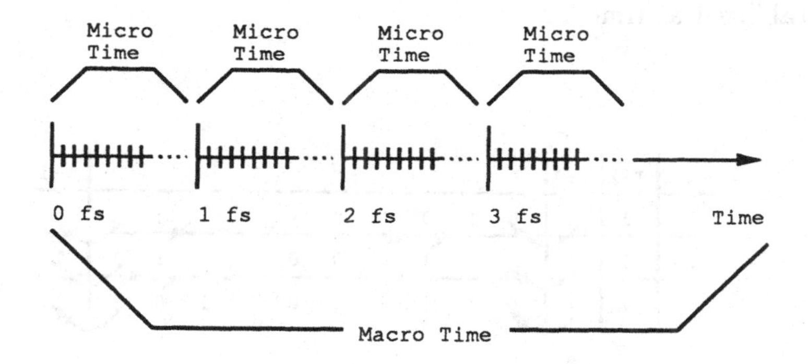

Figure 3.1: Macro and micro time scales in VHDL.

Consider a concurrent signal assignment statement in VHDL:

```
block
  signal X, Y, Z : bit := '0';
begin
  Y <= X;
  Z <= Y;
end;
```

All of the statements in a VHDL block execute in parallel. All languages that allow parallel statements of this form must somehow consistently define the value of `Z` after execution. VHDL assumes a discrete event model of execution. The statement `Y <= X` projects an event on `Y` at some time in the future. That is, the value of `Y` will change to the current value of `X` after some delay. In the case where no delay is specified (as in this example), the amount of delay is one *delta*. Thus the value of Y does not change immediately, but only after one delta. This behavior is illustrated in Figure 3.2. At time T-1 all of the signals have the same value. This becomes their initial value at time T. At time T, assume that some other process in the simulation projects a sequence of events on signal X. When signal X changes value, an event is in turn projected on signal Y. (This is the semantics of the VHDL concurrent signal assignment statement.) A change on Y in turn propagates to signal Z. Macro time advances only when there are no more projected events at time T, i.e., the simulation has stabilized at time T.

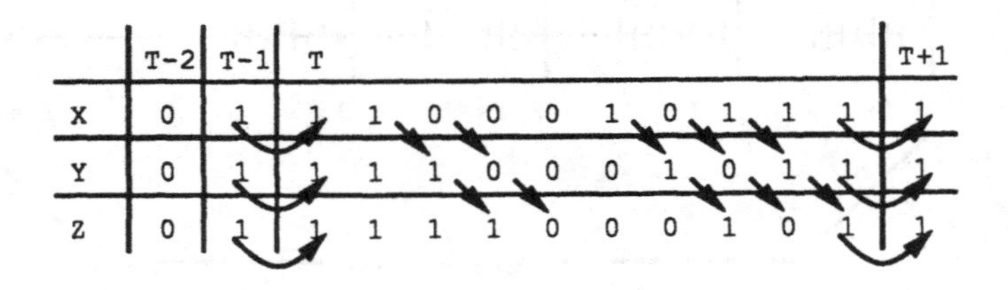

Figure 3.2: Unit delay (delta) time model.

Alternatively, one can understand this execution model by imagining that the expressions on the right-hand side of the assignment statements are all evaluated first (in any order since they have no side effects), and then all the assignments are performed simultaneously. Each evaluate/execute step (simulation cycle) corresponds to one delta in the two-level timing model.

3.1.2 Transport Delay

If a description only uses the unit delay model, then macro time never advances. That is, the model has no notion of real delays. Typically though, a description may specify propagation delays longer than a unit delay. Consider the following VHDL block:

```
block
  signal X, Y, Z : bit := '0';
begin
  Y <= transport X after 1 fs;
  Z <= transport Y after 2 fs;
end;
```

The assignments to signals X and Y take 1 fs and 2 fs, respectively. This means that when X changes value, a change on Y is scheduled for 1 fs in the future. Similarly, when Y changes, a change on Z is projected to occur 2 fs in the future. Note that there are no unit delay steps since a macro time delay has been specified for each assignment.

Delayed signal assignment statements project events (changes in value) on a signal at times greater than or equal to current simulation time. The VHDL simulator maintains a queue of projected events and the times at which they are scheduled. Current time advances when all events scheduled at that time have been executed. When the simulated time advances to a point where a projected change on a signal is scheduled, the value of the signal is actually changed. What happens if more than one change is scheduled on a signal before simulation advances to that time point? The delay model (transport or inertial) determines the behavior in this situation.

This situation is best illustrated by an example derived from [17]. Consider the description of a buffer using transport delay in VHDL:

```
entity Buffer is
  port (input : in bit; output : out bit);
end Buffer;

architecture TransportB of Buffer is
begin
  output <= transport '1' after 10 fs when input = '1' else
                      '0' after 14 fs;
end TransportB;
```

Our intention here is to describe a buffer having a rise time of 10 fs and a fall time of 14 fs. When `input` changes to `'1'` at time T, an event `'1'` is projected on `output` at time `T + 10 fs`. If `input` changes back to

`'0'` t time units later, then an event `'0'` is projected on **output** at time `T + t + 14 fs`. Thus a positive input pulse becomes longer.

When **input** changes to `'0'` at time T, an event (`'0'`) is projected on **output** at time `T + 14 fs`. If **input** changes back to `'1'` t time units later, then an event (`'1'`) is projected on **output** at time `T + t + 10 fs`. Thus a negative input pulse becomes shorter. But what happens if `T + t + 10 fs` < `T + 14 fs`? (That is, t < 4 fs.) The change back to `'1'` will be projected to happen *before* the change to `'0'` ever occurred. In VHDL, the change to `'0'` is canceled. It is *preempted* by the change to `'1'`. Since the output is already `'1'`, no change will be seen at the output.

When an event is projected under the transport delay model, it preempts all events previously scheduled on the signal for that time and all future times. That is, the events are removed from the simulation queue and the simulation behaves as if they had never been projected. We shall refer to this as *forward* preemption. Figure 3.3 shows a sample input waveform and the corresponding output waveform. Arrows from the input to the output waveform indicate the projected and preempted events.

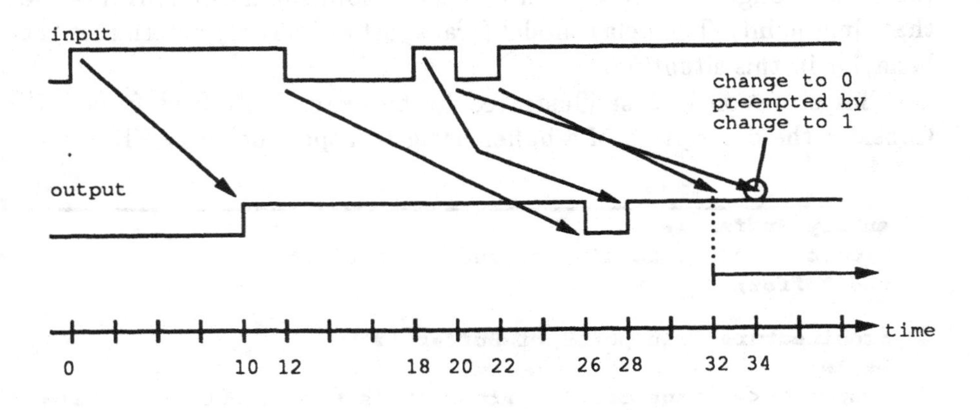

Figure 3.3: Transport Delay Buffer.

3.1.3 Inertial Delay

Transport delay is not the default delay model in VHDL. The default is *inertial* delay. Consider an alternate architecture for the buffer:

```
architecture Inertial of Buffer is
begin
  output <= '1' after 10 fs when input = '1' else
            '0' after 14 fs;
end Inertial;
```

Because inertial delay is the default model in VHDL, thus there is no need for an "inertial" keyword. The sequence of diagrams in Figure 3.4 show how inertial delay operates in the case of the buffer. When `input` changes to `'0'` at time 12, an event is projected at time 26 on `output`. Since there are no other pending events in the event queue for `output`, there is nothing to preempt. At time 18 a change to `'1'` is projected to occur at time 28. The entire event queue is cleared; i.e., the inertial delay model preempts the event queue in *both* directions whereas the transport delay model preempted only future events in the queue. In a similar manner, the change to `'0'` at time 20 preempts the change at time 18, and the change to `'1'` at time 22 preempts the change at time 20. The net result is that no change appears on the output as a result of this sequence of events.

3.1.4 Justification

The need for VHDL inertial and transport delay models may not be readily apparent to the reader. Figure 3.5 shows what happens if there is no preemption (neither transport nor inertial). In this figure, an input transition at time 20 projects an event `'0'` on the output at time 34. A later input transition (at time 22) projects an output change to `'1'` at an earlier time (time 32). In a preemptive model (either inertial or transport) this event would cancel the later projected event. However, without preemption, the effect of this event is lost. The result is that the output waveform goes low at time 34 and remains low, even though the input is high and remains high.

From this example, it is apparent why forward preemption of events, as performed in both inertial and transport delay models, is needed in the

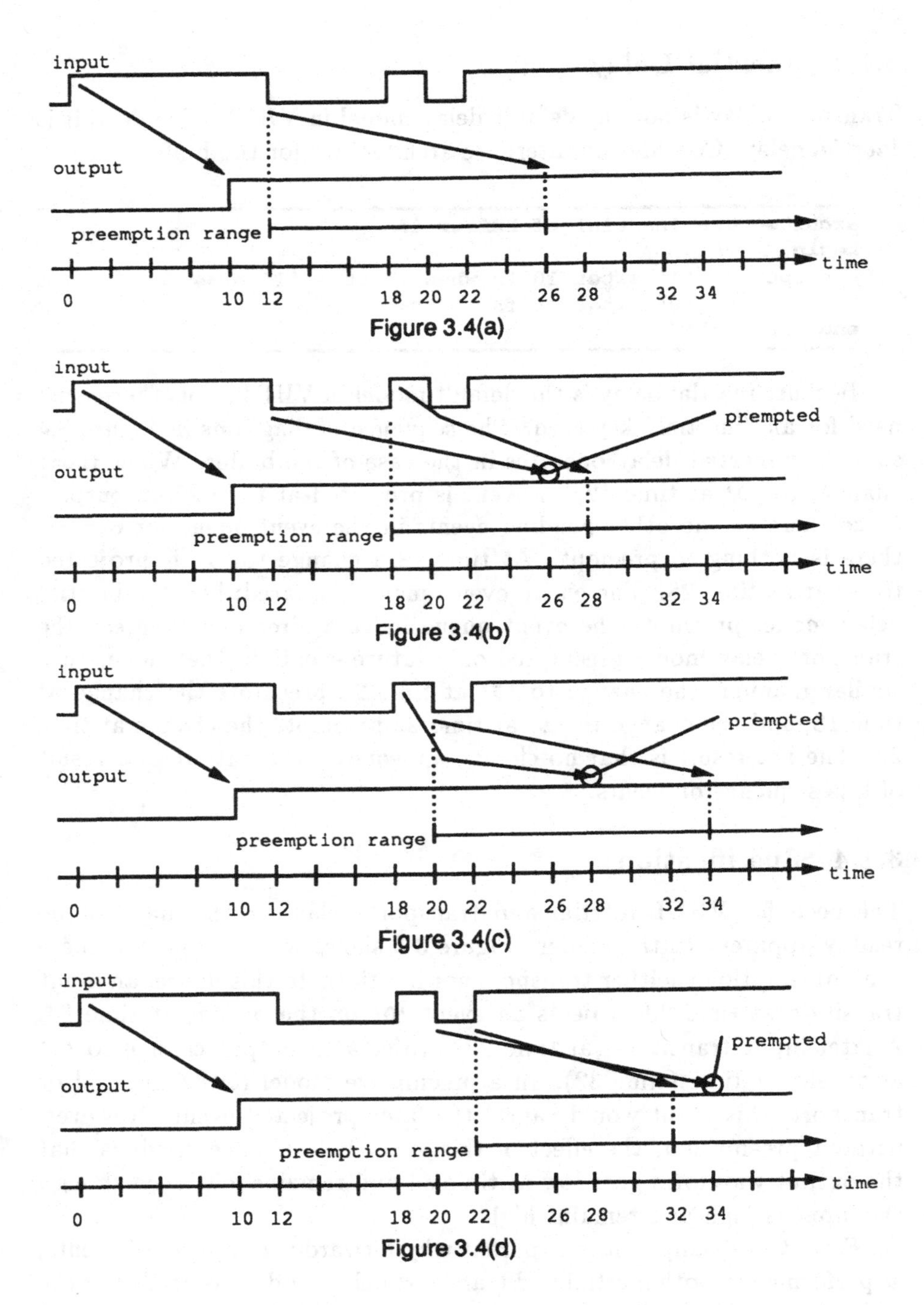

Figure 3.4: Inertial Delay Buffer.

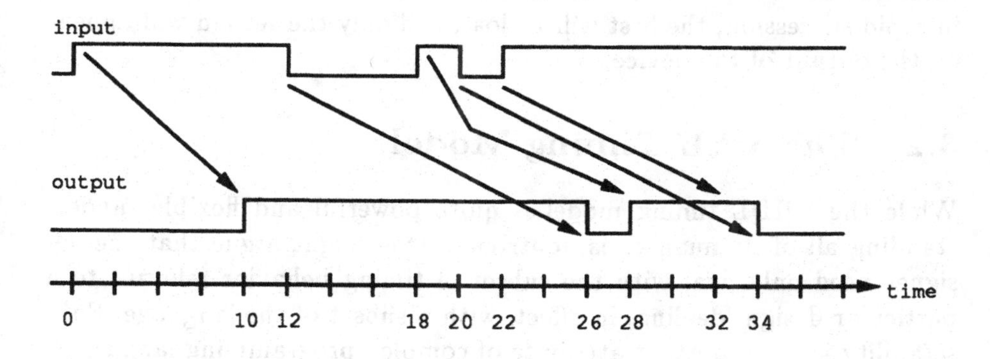

Figure 3.5: Buffer Without Preemption.

VHDL style of description. The difference between inertial and transport models, however, is subtler. Recall that the inertial delay model preempts events in *both* directions in the event queue, whereas transport delay only preempts in the forward direction. The transport delay model describes devices that behave like transmission lines, or have an effectively infinite frequency response. If a device has an output delay of 10 ns, then the transport delay model will permit events at intervals of 1 fs to generate output events delayed by 10 ns at intervals of 1 fs. Multiple events propagate through the device at one time, and thus the device acts like a transmission line.

The inertial delay model describes devices which have a frequency response equal to their delay; i.e. only one event at a time may propagate through the device. Under the inertial delay model, if two pulses occur in rapid succession, the first will be lost, and only the second will appear on the output of the device.

3.2 The VAL Timing Model

While the VHDL timing model is quite powerful and flexible, understanding all of its nuances is nontrivial. One might argue that the designer need only deal with the subset of timing behavior relevant to a particular design, dealing, in effect, with a subset of the language. Subsetability is an important attribute of complex programming languages. Subsetability allows the user to use only a smaller, less complex subset of the language, without fear that more complex features that may not be completely understood will influence the simpler subset. Unfortunately, subsetability is not a substitute for simplicity[12]. The timing model of VHDL is one such example. If the user writes a description without carefully considering the effects of inertial and transport delay semantics on the model, the simulation will most likely not produce the expected results.

In addition, any preemptive model violates two fundamental principles of language design[17]:

1. *The effect of a statement should be a function of only those entities mentioned explicitly in the statement.*

2. ***The effect of a statement should be known upon termination of its execution.***

These principles are important in understanding programs written in any language, and in defining proof rules necessary for verifying their correctness.

To address these issues, VAL uses a single delay model for describing all styles of delay in VHDL. The VAL delay model is *anticipatory*. Since there is no preemption, the user can always be sure of the effect of a statement at the time it is executed. Also, as demonstrated in [1], there are formal methods for analyzing VAL timing specifications. However, VAL introduces timing qualifiers to make it easy to write timing specifications using only constructs which have anticipatory semantics.

3.2.1 Anticipatory Semantics

Behaviors that are described using preemptive semantics can operate on the principle that an event projected now may be canceled at some time in the future. Events that are projected and later canceled were projected without full knowledge of whether or not they should actually occur. At a later time, when more information is available, they are undone.

The principle behind using anticipatory semantics is to wait until all the information needed to determine whether or not an event is to occur is known, and only then project the event. VAL introduces a new language construct (the *timing qualifier*) to make expressing this behavior easier.

Consider the transport delay buffer described earlier in VHDL. The same behavior is modeled in VAL as:

```
when ((input = '0') during[-4 fs, 0 fs]) then
  assert (output[10 fs] = '0');
else
  assert (output[10 fs] = '1');
end when;
```

Because VAL is an annotation language we may only express constraints on **input** and **output**, rather than assigning to the output as is done in VHDL. In the above description, the **during** timing qualifier becomes true when **input** has been low for the past 4 fs. If **input** has been low for at least 4 fs, then **output** will go low 10 fs in the future. Otherwise, **output** will be high 10 fs in the future.

We have explicitly expressed the condition that a negative pulse must be at least 4 fs wide before a change is seen on the output. In the VHDL version this condition is hidden in the semantics of transport delay.

Unlike the VHDL description, the VAL specification does not depend on event driven semantics. The VAL specification is a constraint that should be valid for all time. Given the waveforms for **input** and **output**, the VAL constraint may be checked at any time regardless of the particular sequence of events in the simulator that generated the waveforms.

The specification of the inertial delay buffer in VAL is only slightly more complicated:

```
when ((input = '1') during[-10 fs, 0 fs] then
   (output = '1') during [0 fs, 14 fs];
elsewhen ((input = '0') during[-14 fs, 0 fs] then
   ((output = '0') during [0 fs, 10 fs];
else
   (output = output[-1 fs];
end when;
```

Inertial delay requires that no changes at all occur on the input while a change is propagating through a device. This requires the anticipatory model to check for the stability of the input during the entire propagation delay; 10 fs in the case of a low to high transition and 14 fs in the case of a high to low transition. Note the use of **during** in the assertion about **output**. Once it is known that **input** has been stable high for 10 fs, then **output** must be stable high for 14 fs.

3.2.2 Assertions

A VAL assertion is a process that continuously monitors a boolean expression and reports an error if, under certain conditions, the expression goes false. VAL has four "flavors" of assertions with different temporal properties. The **assert** statement is available in both VAL and VHDL and does not depend on the two-level VHDL timing model. The **finally**,

sometime, and **eventually** assertions are unique to VAL and allow the expression of more powerful assertions that constrain the VHDL two-level timing model in different ways.

Unlike VHDL, assertions in VAL may also be guarded. The assertion is only active when the guard is true. In effect, the assertion never constrains micro cycles where the guard is false. Logically, a guarded assertion such as:

```
when G then
  assert E;
end when;
```

may always be rewritten as:

```
assert E or not G;
```

This is true for all flavors of assertions.

As described previously in Section 2.1.3.3, guarded processes may be nested in VAL. This provides a convenient notation for structuring dependent processes, such as assertions, into hierarchies in which the nesting expresses the dependencies. Without nesting of guards, the user must write a one-level set of guarded processes, e.g., a set of assertions, the guard to each of which contains an appropriate negation of all guards that do not apply when that assertion must be true. Structured specifications using nesting are less likely to contain simple kinds of mistakes.

3.2.2.1 Assert

The VHDL and VAL assertion statements behave in the same manner. They are not sensitive to macro time and check their boolean expression at every delta. If an assertion is guarded, it is checked only if the guard is true in a delta. For example, consider the following VAL assertion:

```
when G then
  assert e;
end when;
```

Figure 3.6 shows the points where a violation occurs as the guard expression (G) and the asserted expression (E) change over time for each flavor of assertion. True and false are represented by "t" and "f" respectively, and a "v" indicates where an assertion violation occurs.

	T-1	T							T+1				T+2			
		1	2	3	4	5	6	7	1	2	3	4	1	2	3	4
G	t	t	f	f	t	t	f	t	t	f	f	t	t	f	f	t
E	t	f	f	f	t	f	f	t	t	t	f	f	f	t	t	f
assert		v				v						v	v			v
eventually						v						v				v
finally												v				v
sometime																v

Figure 3.6: Assertion Violations.

3.2.2.2 Eventually

A common source of errors in hardware design is a signal whose steady-state output is high, but whose intermediate value may change from high to low and back to high before stabilizing. If such a signal controls a latch then the results may not be what is expected.

Such a situation may occur in a VHDL description. The **eventually** assertion is less strict form of **assert** that requires that once a signal becomes true, it must remain true for the remainder of the simulation cycles in the time point. In Figure 3.6, a violation occurs at the fifth delta of time T. The assertion expression **E** became true at the fourth delta, but then went false again in the fifth delta. The eventually process must be active to respond to changes in the assertion expression. Thus no violation is reported in the third delta of T+1 because the guard (**G**) is false.

3.2.2.3 Finally

Most descriptions are only concerned with the value of signals at time points; i.e., at the end of a set of simulation cycles; the particular delta in which the signal changes value is not critical as long as the signal takes on its correct value by the last delta in the time point. In 3.6, a violation occurs at the fourth delta at time T+1. Note that since finally checks only the last delta of a time point, the guard must be true in the last delta so that the finally is enabled.

3.2.2.4 Sometime

The least strict form of checking that VAL permits uses the **sometime** assertion. Sometime requires the assertion expression to be true at some delta in a time point. In Figure 3.6, a violation occurs at the end of time T+2 because the expression E is never true at any delta during that time point when the sometime assertion is active (G is true).

Chapter 4

Designing With Annotations

4.1 Introduction

This chapter illustrates some ways of using VAL annotations in the design process; that is, the process of specifying an entity, and subsequently refining a specification into an architecture of simpler component entities. The top–down approach from specification to implementation is emphasized throughout. The examples given here are intended to be suggestive of various uses of annotations, and are not exhaustive nor complete in any sense. The design process is still very much an area for research, and VAL annotations simply present the designer with a new tool, some of whose applications are shown here.

Generally speaking, the process of developing a design in VAL/VHDL consists of three broad stages:

1. **English Requirements.** The development of an intuitive natural language description of a device.

2. **Specification.** The development of an abstract entity specification expressing the requirements formally in a VAL/VHDL annotated entity declaration.

3. **Implementation.** The construction of a VAL/VHDL architecture body in which VAL annotations specify the components, express

local implementation constraints on the architecture, and define a correspondence (mapping) between the entity abstract state (in the specification) and the states of components in the architecture.

Stage 2 is the process of formalization of requirements. Here, VAL annotations are used to expand a bare VHDL entity interface into an abstract specification of device behavior that expresses all of the given requirements.

The property of being "abstract" is however dependent on technique and ingenuity. Possible abstract specifications at stage 2 may not be equivalent. The design problem in using annotations at stage 2 is to avoid over commitment to particular implementation details. A VAL/VHDL entity interface should specify behavior in such a way that as large a set of implementations as possible satisfies it.

Stage 3 is the process of refining a specification into an architecture consisting of a network of components. Here, VAL annotations are used to express the designer's intentions as to how various facets of the specification are implemented in the architecture. Two examples of this application are mappings and internal constraints.

Mapping annotations (see Section 2.2) express how the abstract entity state is distributed among the states of the components. These annotations specify some of the designer's intentions in the refinement process. They can be an important aid to the designer as the number of components in the network increases.

Internal constraints on a particular network of components are usually needed to ensure that the network will satisfy the specification. These constraints are expressed by VAL annotations in the architecture body.

Annotations should be used as a design aid. They express additional information about the VHDL descriptions (i.e., information lacking in the pure VHDL description) at different levels of detail, and also relationships between those descriptions. They can be reasoned about during the design process, either informally or by formal methods, perhaps thereby influencing the process itself.

The final application is to the automated analysis of designs. Annotations provide comparative simulation (see Section 1.1) whereby the VHDL simulation of an architecture is checked for consistency with an abstract behavior. This facility provides a basis for developing various testing and debugging techniques.

The two examples we will use to illustrate applications of VAL annotations are the Traffic Light Controller (TLC) and the Stack. Our goal in this chapter is to describe how the annotations for the TLC and Stack were written. Later chapters (Chapter 7 for the TLC and Chapter 8 for the Stack) give the complete text of the examples along with simulation results. The complete text for all design units necessary to simulate the examples is not presented here in order to avoid overburdening the reader with excessive detail.

4.2 Traffic Light Controller

Our first example is the Traffic Light Controller (TLC) from [19]. While the TLC is not a particularly large nor complex design problem, it is of sufficient size to illustrate a number of basic concepts while being small enough make a manageable example.

4.2.1 Specification

We first describe the TLC in an English language description adapted from [19]. We then develop a formal VAL/VHDL specification from the English language description.

4.2.1.1 English Language Description

The TLC controls the traffic lights at the intersection of a busy highway and a little-used farm road. Detectors indicate the presence of cars on the farm road waiting to cross or enter the highway. In the absence of any cars on the farm road, the light on the highway should remain green. If a car is detected on the farm road, the highway light is cycled through yellow to red and the farm road light is then to turn green. The farm road light is to remain green only while cars remain on the farm road, but never longer than a specified time limit. The farm road light is then to cycle through caution to red, and the highway light then turns green. Even if cars remain on the farm road, the highway light is to remain green for at least some minimum time limit.

4.2.1.2 VAL/VHDL Specification

We begin by reiterating three goals which we wish to achieve in the VAL/VHDL specification. The specification should satisfy the following three properties:

1. **Consistency**. The specification should be logically consistent, and the timing specification should be causal (see Section 2.1.2).
2. **Generality**. It should be sufficiently general as to permit many different implementations. (It should not overly constrain the implementation.)
3. **Completeness**. It should be sufficiently detailed that any implementation demonstrating undesirable behavior will violate it.

The first goal requires, for example, that assertions about values assumed on inputs or produced by outputs are consistent, and that events specified to occur at one time are not conditions (i. e., guards of processes) for events at a previous time. The second goal has to do with not over committing to a particular implementation. And the third goal involves expressing all of the English requirements.

With these properties in mind, we begin to develop the VAL/VHDL specification.

4.2.1.2.1 Choose types, ports, and timing parameters: Our first step is to choose the types of data flowing across the input and output ports of the entity, the number and type of the ports, and the parameters of the devices timing behavior.

The input and output ports follow from the English language specification. But what should their types be in VHDL? The signal indicating that a car is on the farm road is true or false, so we decide to represent it by a boolean. The color of the highway light may be encoded in many different ways. We decide to postpone the encoding for the moment, and assume a type `COLOR`:

```
entity TLC is
  port (CAR_ON_FARMRD : in boolean;
        HW_LIGHT, FM_LIGHT : out COLOR);
```

```
end TLC;
```

Looking at the English requirements we realize that, at a minimum, we will have three time delays of interest. The first is the length of time which a yellow light lasts. The second is the maximum time that the farm road light may remain green. The third is the minimum time that the highway light must remain green before cycling through yellow to red. In keeping with historical precedent in the design of the TLC, we will assume two time outs; a short time out (STO) controlling the length of a yellow light, and a long time out (LTO) controlling the length of time the farm road light remains green and the minimum time the highway light remains green. Because the absolute value for these parameters is unspecified, we decide to parameterize our design in terms of the two timeout values:

```
entity TLC is
  generic (LTO, STO : TIME);
  port (CAR_ON_FARMRD          : in boolean;
        HW_LIGHT, FM_LIGHT : out COLOR);
end TLC;
```

4.2.1.2.2 Define behavior formally in VAL: Schematically, a clause, or condition, describing a part of the behavior in English is formalized as a guarded assertion process in VAL (see sections 16.3 and 2.1.3.3).

```
when <timed_boolean_expression> then
    <assertion formalizing expected behavior> else
    <report error in observation>
```

Now we translate the English language requirements into VAL annotations.

1. When there is a car on the farm road, and the highway light has been green for at least LTO, then cycle the highway light to yellow.

```
--| when (CAR_ON_FARMRD and
--|       (HW_LIGHT = GREEN) during LTO) then
--|   finally HW_LIGHT'changed(YELLOW)
--|     else report "Highway light did not change " &
--|                 "from green to yellow";
--| end when;
```

Before proceeding further, let us examine this annotation carefully to make sure it says what we want. The expression `((HW_LIGHT = GREEN) during LTO)` is a boolean expression that will become true when the signal `HW_LIGHT` has been green for a time interval `LTO` preceding the current time. If this expression is true, and there is also a car on the farm road, then the highway light should change color to yellow.

The problem with this annotation is that it overly constrains the specification. We don't mean to say that the highway light must change color immediately, but rather that it must change color within some short delay of when the guard conditions become true. Note that our concern here is not with how many deltas (simulation cycles) the change requires to take place. The finally flavor of assertion will assure that the value checked is the last stable value at the end of a time point.

Rather our concern is that even the fastest implementation will require a few milliseconds to react and cause the highway light to change color. The above annotation does not allow for the reaction time of the implementation. Thus the annotation should read, "If this expression is true, and there is also a car on the farm road, then the highway light should change color to yellow *within* some small delay."

Assuming a generic parameter `SLOP` of type `TIME` representing the small delay, the annotation can be rewritten as:

```
--| when (CAR_ON_FARMRD and
--|       (HW_LIGHT = GREEN) during LTO) then
--|   finally HW_LIGHT'changed(YELLOW) within [0 ns, SLOP]
--|     else report "Highway light did not change " &
```

```
--|                    "from green to yellow on time";
--| end when;
```

In general an English language expression of the form "signal s within time t" can be written in VAL as "s within [0,t]." Section 2.1.4.4 explains how to write such "loose" timing constraints.

2. When the highway light is yellow, it should only stay yellow for STO before turning red. The farm road light should turn green at the same time.

```
--| when ((HW_LIGHT = YELLOW) during STO) then
--|    finally HW_LIGHT'changed(RED)) within [0 ns, SLOP]
--|      else report "Highway light did not change " &
--|                  "from yellow to red on time";
--|    finally FM_LIGHT'changed(GREEN) within [0 ns, SLOP]
--|      else report "Farm road light did not change " &
--|                  "to green on time";
--| end when;
```

3. Cycle the farm road light to yellow when the farm road light is green, and there are no cars on the farm road, or when the farm road light is green and has been green for LTO.

```
--| when ((FM_LIGHT = GREEN) and (not CAR_ON_FARMRD)) or
--|       ((FM_LIGHT = GREEN) during LTO) then
--|    finally FM_LIGHT'changed(YELLOW) within [0 ns, SLOP]
--|      else report "Farm road light did not change " &
--|                  "to yellow on time";
--| end when;
```

4. When the farm road light has been yellow for STO, cycle it to red. The highway light should turn green at the same time.

```
--| when ((FM_LIGHT = YELLOW) during STO) then
--|    finally FM_LIGHT'changed(RED) within [0 ns, SLOP]
```

```
--|     else report "Farm road light did not change " &
--|                 "from yellow to red on time";
--|   finally HW_LIGHT'changed(GREEN) within [0 ns, SLOP]
--|     else report "Highway light did not change " &
--|                 "to green on time";
--| end when;
```

5. Implied in the English specification is the idea that the farm road light and the highway light should never both be green. There are two ways to deal with this implied property. It can be expressed as an assertion as below. Or we can prove that it is a consequence of the VAL specification, in which case the assertion is unnecessary.

```
--| assert not ((FM_LIGHT = GREEN) and
--|             (HW_LIGHT = GREEN))
--|      else report "Danger - Farm road and highway " &
--|                  "lights are both green";
```

All of the annotations up to this point have specified when a change must occur. However, they have not specified that these are the *only* changes that may occur. For example, the highway light may change from red to green at some random time, and the specification we have so far will allow it. Constraints must be added that specify that when a change in color of a light occurs, that the prerequisites for that change were met. Because these constraints are very similar to those we have described already, we will not enumerate them here. The reader is referred to Chapter 7 for the complete listing of the TLC specification including these constraints.

4.2.1.3 Commentary

The two basic steps of choosing the ports, data types, and timing parameters, and formalizing the behavioral specification may need to be iterated. That is, we may not declare all of the types or timing parameters at first, but may uncover some that were missed when we initially formalized the behavior. We alluded to this when we modified the TLC description to include the SLOP timing parameter. Also, some generic parameters may arise after the behavior is formalized, when we try to express it in a more general (generic) form.

Notice that the final behavioral specification of the TLC consists of a set of concurrent guarded processes and a concurrent unguarded assertion process. The assert process is checked concurrently with all of the other guarded processes. The guards, however, are intended to be mutually exclusive, and we should probably add additional assertions to verify this condition, unless we can prove that mutual exclusion is a consequence of the specification.

Finally, the VAL specification is easy to reason about. For example, assume that initially both lights are not `GREEN`. Assuming no other changes of lights occur than the ones specified, it is easy to reason informally that both lights can never be `GREEN` at the same time. Without this assumption, the illegal combination may be produced by an implementation that satisfies the behavioral specification but not the assertion. An illegal combination would have to occur during the sloppy intervals within which changes are specified to occur. So the assertion is necessary.

4.2.2 Implementation

As mentioned previously, architecture annotations can be used to specify a mapping between the abstract state of an entity and the abstract states of architecture components, and also to specify constraints on the architecture. We present here an implementation for the TLC and give examples of these applications.

Let us say first, however, that we do not attempt to give rigorous methods for refining VAL specifications into architectures. This area, automated synthesis, is an active research area. Rather, we assume that refinement of abstract specifications into detailed architectures is achieved by some as yet to be defined process. Certainly such processes may be aided by annotations. But, given an architecture, it is now our task to express formally as much information as possible about the architecture and its relationship to the abstract specification. Figure 4.1 shows the original (prior to the addition of VAL annotations) architecture.

The implementation contains undocumented design decisions that we should annotate both for readability and for an extra degree of run time

```
architecture A of TLC is
    constant CYCLE_TIME : TIME := 100 ns;
    signal QHG, DHG : bit := '1';
    signal QFG, DFG, QFY, DFY, QHY, DHY : bit := '0';
    signal ST, LT, CLK : bit:= '0';
    component dff_c
        port(D, CK : in bit; Q : out bit);
    end component;
begin
    SHY : dff_c port map (DHY, CLK, QHY);
    SFG : dff_c port map (DFG, CLK, QFG);
    SFY : dff_c port map (DFY, CLK, QFY);
    SHG : dff_c port map (DHG, CLK, QHG);
    CLK <= not CLK after (CYCLE_TIME / 2);
    DHY <= '1' when ((QHG = '1') and (LT = '1') and
                      (CAR_ON_FARMRD)) or
                     ((QHY = '1') and not (ST = '1'))
                else '0';
    DHG <= '1' when ((QFY = '1') and (ST = '1')) or
                     ((QHG = '1') and not (CAR_ON_FARMRD and
                                          LT = '1')) or
                     ((QHY = '0') and (QFG = '0') and
                      (QFY = '0') and (QHG = '0')) else '0';
    DFY <= '1' when ((QFG = '1') and
                     ((LT = '1') or (not CAR_ON_FARMRD))) or
                     ((QFY = '1') and not (ST = '1'))
                else '0';
    DFG <= '1' when ((QHY = '1') and (ST = '1')) or
                     ((QFG = '1') and CAR_ON_FARMRD and
                      (LT = '0')) else '0';
    LT <= '1' after LTO when ((QFG = '1') or (QHG = '1'))
            else '0' after 1 fs;
    ST <= '1' after STO when ((QFY = '1') or (QHY = '1'))
            else '0' after 1fs;
    HW_LIGHT <= GREEN when QHG = '1' else
                YELLOW when  QHY = '1' else RED;
    FM_LIGHT <= GREEN when QFG = '1' else
                YELLOW when  QFY = '1' else RED;
end A;
```

Figure 4.1: TLC Original Architecture

checking.

For example, the designer has chosen to use a one-hot encoding for representing the state. This implies that exactly one of the outputs from the D flip-flops must always be asserted.

We can express this by enumerating all the legal combinations of **QHY**, **QFG**, **QFY**, **QHG**.

```
--| finally (((QHY = '1') and (QFG = '0') and
--|          (QFY = '0') and (QHG = '0')) or
--|         ((QHY = '0') and (QFG = '1') and
--|          (QFY = '0') and (QHG = '0')) or
--|         ((QHY = '0') and (QFG = '0') and
--|          (QFY = '1') and (QHG = '0')) or
--|         ((QHY = '0') and (QFG = '0') and
--|          (QFY = '0') and (QHG = '1')))
--|    else report "Conflicting value for state";
```

4.2.2.1 Commentary

The difficult part of annotating an architecture is managing to not *over-annotate*. For example, it is tempting to describe the expected behavior of the flip-flops in the TLC architecture. However, we must realize that the specification for the flip-flops appears in their entity specification, and that it is not our responsibility to verify their behavior in the TLC architecture.

An excellent application for architecture annotations is the checking of invariants implied by the logic in the architecture. The assertion above specifies that exactly one of **QHY**, **QFG**, **QFY**, and **QHG** is true at any time. If there are wiring errors in the architecture, or errors in the logic, this assertion may catch the error.

What we have not done is simply repeat the completely functionality of the architecture in assertions. Instead, we have added a simple *additional* check on the behavior. For example, we could have written the logic for the next state variables in VAL and checked it against the VHDL behavior. But while this would certainly give a greater degree of consistency checking, it is tedious, and it is unreasonable to expect a designer to maintain two similar descriptions of the same device at approximately the same level of abstraction.

4.3 Stack

While the TLC example demonstrates a number of useful concepts in VAL, it does not demonstrate the use of the entity state model. That is, we did not find it necessary to maintain any state information as part of the specification of the TLC. In effect, the "state" of the TLC was known completely from its ports. This is not the case with the Stack. The Stack will require the use of an abstract type to maintain state information.

4.3.1 Specification

As with the TLC, we present first an English language description of the Stack, and then proceed to develop a formal VAL/VHDL specification from it.

4.3.1.1 English Language Description

The Stack is a device which stores data in a last-in, first-out manner. When the Stack is enabled, and a push operation is indicated, a clock signal causes the stack to store the data appearing on its input port. The value most recently stored (the top of the stack) is always available on the output. The Stack provides a signal to indicate that it is full, and ignores all push operations when it is full.

When the stack is enabled, and a pop operation is indicated, a clock signal causes the most recently stored element (the top of the stack) to be discarded, and the next most recently stored element to become available on the output port. If the stack is empty, a signal is asserted to indicate that there are no more elements, and we don't care what appears on the output port.

4.3.1.2 VAL/VHDL Specification

In order to model the Stack in VAL, we will need some software support for modeling its state. Recall that VAL is intended to describe the relative timing relationships between events that occur in the hardware. VAL is not intended for performing the *computation* associated with those events. That is, VAL provides new constructs for indicating *when* an output will take on a new value, but it relies on constructs of VHDL or other programming language interfaced to VHDL for computing (in complex cases) the new value of the output. Here, we use the procedural portion of VHDL to express the abstract stack state type.

So we begin by developing a VHDL package designed to manipulate stacks. Note that any programming language with an interface to VHDL would be suitable for developing this package. [1] The lack of abstraction mechanisms in VHDL, such as generic packages and privates types as found in Ada, limits somewhat our ability to develop a reusable stack package. However, since we are using VHDL, we have tried to keep the example as independent of the element type and maximum depth as possible. The element type is declared in its own package:

```
package Element is
  subtype ElementT is bit_vector(0 to 3);
  constant NullElement : ElementT :=
                         ElementT'('0','0','0','0');
end Element;
```

This package is then imported by the VHDL stack package shown in Figure 4.2. For concreteness, we have chosen to make this a stack of 4-bit elements, with a maximum depth of 5 elements. In a language supporting generics (such as Ada), the depth and element types would be generic parameters.

A great deal of time and effort may be spent creating packages such as **StackP**. However, a good design environment would incorporate a library of such packages allowing the designer to simply select from one of many available abstractions for modeling the device. (e.g., stacks, FIFO queues, memories, etc.) The particular implementation (package body) we will eventually use for this package is not important to us here. One possible implementation in given in Chapter 8.

4.3.1.2.1 Choose types, ports, and timing parameters: From the English requirements, we choose to specify separate input and output

[1] Although VHDL does not define a foreign language interface, and current simulators do not yet support foreign language interfaces, foreign language interfaces are planned for simulators currently under development. Such interfaces will be, at least initially, simulator dependent.

```
use work.Element.all;
package StackP is
  constant depth : natural := 5;
  type stackarray is array(natural range 1 to depth) of
                                                  ElementT;
  type StackT is record
    top : natural;
    stack : stackarray;
  end record;
  constant EmptyStack : StackT :=
    StackT'(0, StackArray'(others => NullElement));
  function SPush(s : in StackT; e : ElementT)
               return StackT;
  function SPop(s : in StackT) return StackT;
  function STop(s : StackT) return ElementT;
  function is_Full(s : StackT) return boolean;
  function is_Empty(s : StackT) return boolean;
end StackP;
```

Figure 4.2: VHDL Stack Package

ports, a direction port indicating a push or pop, a clock port, and an enable port. The type of the data ports is determined by the type of the elements stored in the Stack. Using the **Element** package just developed to help in modeling the Stack, the data ports will be of type **ElementT**. The VHDL entity declaration is thus:

```
use work.Element.all;
entity Stack is
    port(din             : in ElementT;
         dout            : out ElementT;
         dir, clk, ce    : in bit;
         full, empty     : out bit);
end Stack;
```

Clearly there are numerous timing parameters that we could specify; the delay from a pop or push to an output change, setup and hold times for a push, the delay for full or empty to change, etc. The timing parameters are specified in the same manner as they were for the TLC. Because our focus in this example is on the use of the state model, and the specification of timing would tend to distract from that goal, we choose to neglect timing for the moment in this example. However, we emphasize that in order to be useful as a specification for a real hardware device timing information will have to be added.

4.3.1.2.2 Define behavior formally in VAL: Recall that in Chapter 1 we said that VAL should not be concerned with the details of how a computation is performed, but rather with the specification of *what* computation is performed *when*. Using the stack package, we now formalize the English description in VAL by specifying *what* operations of the abstract data type are performed *when*. First we must declare that the device state will be modeled by a stack. The package **StackP** is imported into the entity specification, and the state model is declared as the **StackT** type exported from **StackP**:

```
--| state model is StackT := EmptyStack;
```

Now we can translate the English description into a set of VAL constraints.

1. When there is a rising edge on the clock, and the stack is enabled, then the stack may take some action. If the direction indicates a push, and the stack is not full, then take the element on the input port and push it on the stack. If the direction indicates a pop, and the stack is not empty, discard the top element from the stack.

```
--| when (clk'changed('1') and (ce = '1')) then
--|      when (dir = '1') then
--|           when not (is_Full(state)) then
--|                state <- SPush(state,din);
--|           else
--|                assert false
--|                  report "push attempted on full stack";
--|           end when;
--|      else
--|           when (not is_Empty(state)) then
--|                state <- SPop(state);
```

```
--|             else
--|                 assert false
--|                     report "pop attempted on empty stack";
--|             end when;
--|         end when;
--|     end when;
```

2. The top of the stack always appears on the output port if the stack is not empty.

```
--| when(not is_Empty(state)) then
--|     finally (dout = STop(state));
--| end when;
```

Notice that we have not said anything about the value on the output port when the stack is empty.

3. An output indicates when the stack is empty.

```
--| when (is_empty(state)) then
--|     finally(empty = '1');
--| else
--|     finally(empty = '0');
--| end when;
```

4. An output indicates when the stack is full.

```
--| when (is_full(state)) then
--|     finally(full = '1');
--| else
--|     finally(full = '0');
--| end when;
```

The complete annotated entity interface combining all the annotations appears in Chapter 8.

4.3.1.3 Commentary

Notice that the VAL stack entity specification describes *when* values are available on out ports in response to values on in ports. That data types and operations encapsulated in the **StackP** package are used, but their computational details are hidden in the package body for **StackP**. We may assume that the VAL stack specification is executable by assuming a body for **StackP**. One possible body appears in Chapter 8. This allows comparative simulation of this specification against a VHDL architecture.

If we wish to apply formal proof methods to the VAL stack, we would need the **StackP** package specification to contain an abstract algebraic specification of its operations. Since we are describing a hardware system, we could now add timing information to the specification to indicate a range of times over which the computations must be performed.

4.3.2 Implementation

Once again, we will not attempt to describe the design process in which an implementation is constructed from a specification. Instead, we will simply present an implementation, and give some insight into it through the addition of annotations.

An implementation for the Stack is shown in Figure 4.3. The Stack is constructed from a set of registers, connected using some logic and multiplexors. A diagram of the structure is shown in Figure 4.4. As with the TLC there are a number of undocumented design decisions that can be made explicit using annotations.

```
architecture structure of Stack is
    type ElementVector is array(Natural range 1 to depth) of
                                                    ElementT;
    signal zero : bit := '0';
    signal one : bit := '1';
    component creg
        port(din : in ElementT;
             dout : out ElementT;
             en : in bit;
             ck : in bit);
    end component;
    component cmux
```

```
            port(d1, d2 : in ElementT;
                 dout : out ElementT;
                 s : in bit);
        end component;
        component shiftreg
            generic(size : natural);
            port(lin : in bit;
                 rin : in bit;
                 dout : out bit_vector(1 to size);
                 shift : in bit;
                 dir : in bit);
        end component;
        signal mux0in : ElementVector;
        signal mux1in : ElementVector;
        signal muxout : ElementVector;
        signal regout : ElementVector;
        signal srout : bit_vector(1 to depth);
        signal enable, sshift : bit;
    begin
        G1: for i in 1 to depth generate
            M1 : cmux
                port map(mux0in(i), mux1in(i), muxout(i), dir);
            R1 : creg
                port map(muxout(i), regout(i), enable, clk);
            G2 : if (i >= 2) generate
                mux0in(i - 1) <= regout(i);
                mux1in(i) <= regout(i - 1);
            end generate;
        end generate;
        mux1in(1) <= din;
        mux0in(depth) <= NullElement;
        SR : shiftreg
            generic map(depth)
            port map(one, zero, srout, sshift, dir);
        empty <= not srout(1);
        full <= srout(depth);
        enable <= ce and ((dir and (not srout(depth))) or
                  ((not dir) and srout(1)));
        sshift <= ce and (enable and clk);
        dout <= regout(1);
    end structure;
```

Figure 4.3: Original Stack Architecture

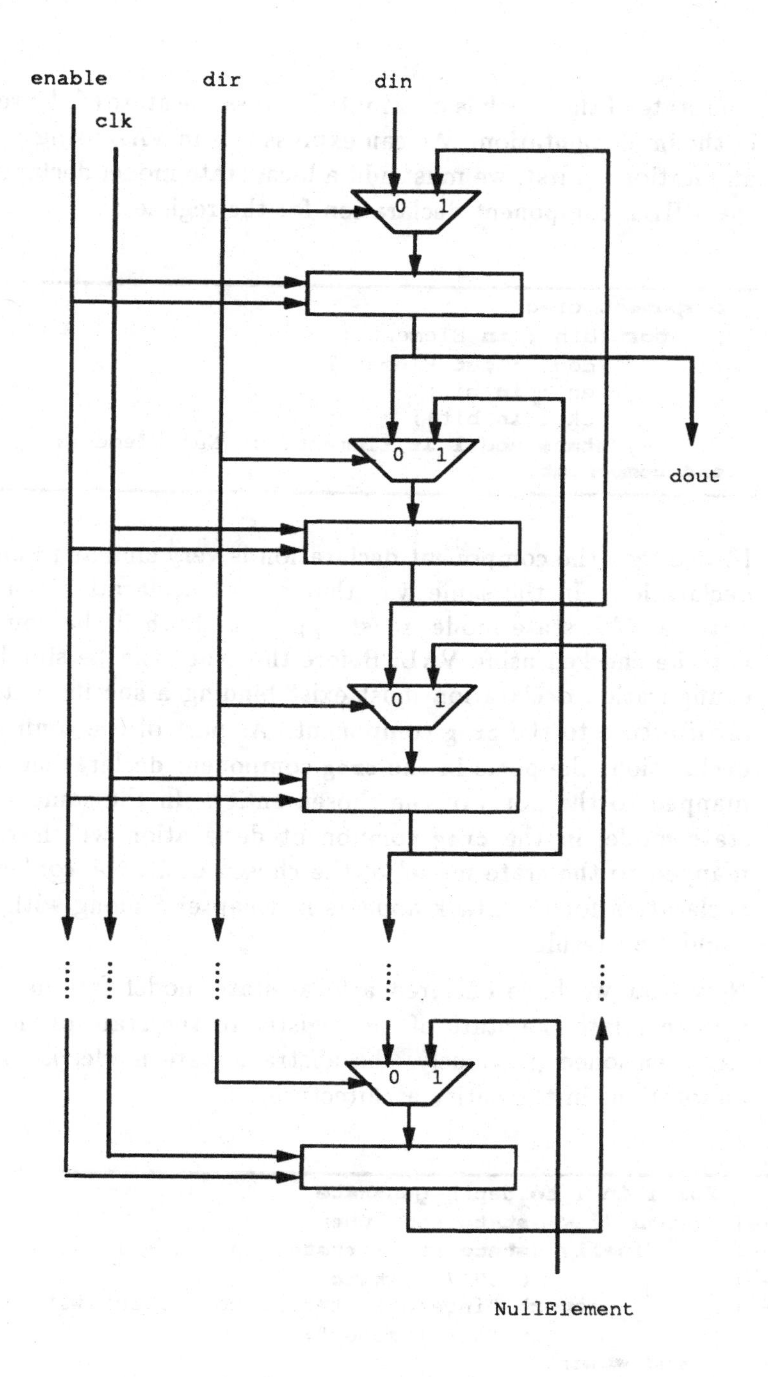

Figure 4.4: Diagram of Stack Data Path

1. The state of the Stack is distributed across the state of the registers in the implementation. We can express this in VAL using mapping annotations. First, we must add a local state model declaration to the VHDL component declaration for the register:

```
component creg
    port(din  : in ElementT;
         dout  : out ElementT;
         en : in bit;
         ck : in bit);
    --| state model is ElementT := NullElement;
end component;
```

Notice that the component declaration is very similar to an entity declaration. In the same way that a port declaration appears in both, a VAL state model must appear in both if the component is to be checked using VAL. Before the Stack can be simulated, a configuration declaration must exist binding a specific entity and architecture to the **creg** component. As part of the configuration declaration, the ports in the **creg** component declaration must be mapped to the ports of the chosen entity. In the same way, the state model in the **creg** component declaration will have to be mapped to the state model of the chosen entity. A configuration declaration for the Stack appears in Chapter 8 along with sample simulation results.

Now that we have declared a local state model for the register, we can relate the state of the register to the state of the Stack. (As mentioned previously, the abstract state model is visible to annotations in the entity architecture.)

```
--| for i in 1 to depth generate
--|   when (i <= state.top) then
--|     finally (state.stack(state.top + 1 - i) =
--|              G1.R1(i).state)
--|         report "internal state inconsistent with val " &
--|                "state model";
--|   end when;
```

```
--| end generate;
```

Note that **state.top** refers to the top component of the record, StackT, that is the state model. Also, **G1.R1(i)state** refers to the state model of the ith register.

This VAL statement generates a set of assertions, one for each of the registers, that checks if the state of the register is equal to the ith data element stored in the state of the entity. Equivalence is checked only for entries that are currently on the stack; the "empty" registers are not checked.

2. The implementation uses the "Johnson Counter" technique for maintaining a count of the number of elements on the stack. A shift register acts as a counter. The shift register is shifted up when a push occurs and down when a pop occurs. Ones are always shifted into the bottom of the shift register on a push, and zeros are shifted into the shift register on a pop. When the Stack is full, all the bits of the shift register are ones. When the Stack is empty, all the bits of the shift register are zero. An intermediate state is characterized by the lower portion of the shift register containing all ones and the upper portion containing all zeros. The number of ones will indicate the number of elements on the Stack.

   We can make this design decision explicit by adding a VAL annotation relating the abstract state model of the entity declaration to the outputs of the shift register:

```
--| for i in 1 to depth generate
--|   when (i <= state.top) then
--|     finally srout(i) = '1'
--|         report "srout state inconsistent with val " &
--|                "state model: 1 expected";
--|   else
--|     finally srout(i) = '0'
--|         report "srout state inconsistent with val " &
--|                "state model: 0 expected";
--|   end when;
--| end generate;
```

The complete text of the Stack implementation, including the annotations, appears in Chapter 8.

4.3.2.1 Commentary

The first architecture annotation is an example of a mapping annotation. The abstract state model of the stack is related to the abstract state models of the components of the stack (the registers). Since the state of each register is equivalent to the data visible on its outputs, this annotation could also have been expressed by relating the state of the stack to the value appearing on the registers' outputs. This approach was taken in the second annotation, where the abstract state model of the stack is related to the output of the shift register. In this example the components happen to have the property that their state is always visible on their outputs, allowing us to write either annotation. However, in general the state of the components may not be completely visible at the outputs, and in that case mapping annotations are stronger constraints than annotations that simply constrain the ports of components.

Notice that both annotations use a **when** statement within the generate. Because a generate statement is restricted to a constant range, we must use a guard to control the activation of the assertion. In general, whenever we want to make an assertion about a dynamically varying range of signals or components in an architecture, we have to use a generate to loop over all the signals or components in the set, and use guards to select just those in the range of interest.

As in the TLC, the annotations appearing in the architecture do not simply repeat the behavior of the architecture or the components in the architecture. Instead they assert some internal consistency checks that the designer knows must be true. In the specific case of the stack, the annotations particularly help in explaining how the storage in the stack specification is implemented in the architecture. The first annotation tells us that the elements in the stack are stored in the registers, with the top element on the stack always appearing in the first register, the second element always appearing in the second register, etc. The second annotation tells us that if there are N elements on the stack, then the first N bits of the shift register must be 1, and the remainder must be 0. Hopefully, the actual behavior of the components combined with the way they have been connected in the architecture will guarantee that these annotations always hold.

4.4 Summary

This chapter presented the design of two devices using annotations. In each case the English language description of the device was given first, and then translated into a formal specification in VAL/VHDL. Once the device has been designed, the formal specification can be used during simulation to verify that the result of simulating the architecture satisfies the formal specification.

The process of specifying and designing hardware is normally not a simple top-down process. Typically both the specification and design will change as the requirements are better understood during the course of the design. Also, the process of formalizing the requirements in VAL/VHDL will cause the requirements to change as their implications become apparent. For example, our attempt to formalize the specification of the TLC lead to the introduction of timing slop into the requirements and specification.

Constructing specifications that are consistent, general, and complete is not a trivial undertaking and requires some experience in using VAL annotations.

The process of constructing an architecture to satisfy a specification should use annotations to express design decisions (e.g., the one-hot encoding and Johnson Counter) explicitly. This should be done when such decisions are first made, and not as an afterthought. Mapping annotations are particularly useful, as demonstrated by the Stack, for explaining what piece(s) of an architecture implement some component of the entity state.

In annotating an architecture it is important not to over-annotate the design. Since formal specifications for the device and its components already exist, annotations in the architecture that simply repeat this information are unnecessary. Useful architecture annotations will constrain internal aspects of the architecture such as connectivity that may introduce errors into the design even if each of the components of the architecture behaves as expected. State provides a useful internal checking point for the architecture. If the internal state of the architecture becomes inconsistent with the state of the entity, yet each of the components is behaving correctly, then the scope of the error is limited to the connection of those components.

We hope the examples presented here have served to illustrate some useful techniques in specification construction and architecture annotation. Part II of this book contains a number of examples for further reference, including the complete text of the TLC and Stack examples.

Part II

Examples

Chapter 5

Crazy AND Gate

5.1 Requirements

The AND gate in this example has different propagation delays from each of its two inputs. In addition, the propagation delay depends on the current output of the gate. Such delay characteristics are not unusual in real implementations of AND gates.

The AND gate has two inputs `X` and `Y`, and one output `Z`. When the output transition is caused by the change of input `X`, the propagation delay for high to low change at output `Z` is 14 fs, and that for low to high transition is 18 fs. When the output transition is caused by the change of input `Y`, the propagation delay is 15 fs for high to low transition at `Z`, and 20 fs for low to high transition. In modeling the timing behavior of the AND gate, we assume the propagation delay occurs first, and then the actual AND operation occurs.

In summary:

X	Y	Z	Delay
1	$1 \rightarrow 0$	$1 \rightarrow 0$	15 fs
1	$0 \rightarrow 1$	$0 \rightarrow 1$	20 fs
$1 \rightarrow 0$	1	$1 \rightarrow 0$	14 fs
$0 \rightarrow 1$	1	$0 \rightarrow 1$	18 fs

5.2 Entity Declaration

After considering the description of the timing behavior of the device, we note that Y must stay high for at least 5 fs (20 fs − 15 fs) for the output to go high. Similarly, the X input must stay high for at least 4 fs (18 fs − 14 fs) for the output to go high. Since this is an AND gate, both conditions must hold for the output to actually go high. Otherwise, the output is low. So we have the following VAL/VHDL entity declaration:

```
-- Annotated entity declaration
entity CRAZY_AND is
  port (X, Y : in BIT; Z : out BIT);
begin
  --| when (Y = '1' during 5 fs)[-15 fs] and
  --|      (X = '1' during 4 fs)[-14 fs] then
  --|    finally Z = '1';
  --| else
  --|    finally Z = '0';
  --| end when;
end CRAZY_AND;
```

5.3 Commentary

This simple example demonstrates the expressiveness of VAL timing constructs. When using VHDL, one must inevitably introduce state variables in describing this example; consequently, the description treats the Crazy AND as a sequential logic device although and AND gate is generally considered a combinational logic device.

Consider a VHDL architecture potentially satisfying the specification:

```
-- VHDL architecture
architecture A of CRAZY_AND is
    signal X1, Y1 : bit;
begin
    X1 <= '1' after 18 fs when X = '1' else
          '0' after 14 fs;
    Y1 <= '1' after 20 fs when Y = '1' else
          '0' after 15 fs;
```

```
        Z <= X1 and Y1;
    end A;
```

The architecture declares two signals (**X1** and **Y1**) that hold delayed values of the inputs. The assignment statements to those signals read almost exactly as the informal specification of the device. The VAL Transformer can be used to verify the behavior of the VHDL description. In order to simulate the **CRAZY_AND**, we construct the following test bench and associated configuration:

```
entity tb is
end tb;

architecture a of tb is
    component andc
        port (X, Y : in BIT; Z : out BIT);
    end component;
    signal in1, in2, output : bit;
begin
    c1 : andc
        port map(in1, in2, output);
    in1 <= '1',
           '0' after 250 fs,
           '1' after 280 fs;
    in2 <= '0',
           '1' after 100 fs,
           '0' after 103 fs,
           '1' after 200 fs,
           '0' after 210 fs,
           '1' after 220 fs,
           '0' after 310 fs;
end a;

use work.all;
configuration c of tb is
  for a
    for c1 : andc
        use entity CRAZY_AND(A);
        --| valentity;
    end for;
  end for;
end c;
```

The VAL Transformer is first run with all of the above VHDL descriptions as input. It produces a new VHDL description in which the VAL annotations have been replaced by VHDL code to perform the checking, and a script that issues the proper commands to compile and run the simulation. (Where appropriate, all examples in this book have been simulated on a Sun-3/110 running version 2.1 of the Intermetrics standard VHDL 1076 Support Environment.)

The simulator produces the following simulation results:

```
TIME  |--------------------SIGNAL NAMES------------------|
      |
(FS)  |           IN1              IN2            OUTPUT
      |
   0  |           '0'              '0'            '0'
  +1  |           '1'              '0'            '0'
 100  |                            '1'
 103  |                            '0'
 200  |                            '1'
 210  |                            '0'
 220  |                            '1'
 221  | %VHDSIM-W-ASSERTV Assertion Violation after 221 fs
      |  at line 132 in design unit <<LMA.BOOK.VALAND>>VALO
      | UT_CRAZY_AND(STATEMONITOR)
      | Assertion violation
 240  |
  +2  |                                           '1'
 250  |           '0'
 264  |
  +2  |                                           '0'
 280  |           '1'
 298  |
  +2  |                                           '1'
 310  |                            '0'
 325  |
  +2  |                                           '0'
```

The assertion violation indicates that something is wrong with the VHDL description. Examining the inputs immediately prior to the violation, we see that **IN1** has been high since the start of the simulation, and that **IN2** changed from `'0'` to `'1'` at 200 fs, from `'1'` to `'0'` at 210 fs, and from `'0'` to `'1'` at 220 fs. In the test bench architecture, **IN1** is connected to the **X** input and **IN2** is connected to the **Y** input. From the previous table, a change from `'0'` to `'1'` on **Y** should produce a change to `'1'` on the output after 20 fs, if no other faster change occurs before then.

since the next input change doesn't occur until 210 fs, the output should change to `'1'` at 220 fs. When this doesn't happen, VAL generates an assertion violation. In general, VAL generates an assertion violation as soon as it is possible to detect the violation. In this case the violation is detected at the next smallest time point, 221 fs.

The source of the error in the VHDL description is a subtle one that is often overlooked by writers of VHDL modules. The signal assignment statements in the VHDL description use inertial delay for placing events in the event queue. This means that all events are cleared from the queue when an event is placed in the queue. Chapter 3 describes the VHDL timing model in more detail. In this case, the change on **IN2** at 210 fs preempts the change scheduled earlier at 200 fs. The result is that the effect of the event at 200 fs is lost.

We now have two choices: (1) we can change the specification of the device, or (2) we can change the implementation. It is educational to explore both possibilities.

5.3.1 Altering the Specification

In order to change the specification, we must fully understand what is happening with the VHDL inertial delay timing model (Chapter 3). A simplified description of inertial delay is that the signal must stay stable for at least the length of its propagation delay for an change to have an effect.

Using this technique produces the following VAL specification:

```
-- Annotated entity declaration
entity CRAZY_AND is
  port (X, Y : in BIT; Z : out BIT);
begin
  --| when (Y = '1' during [-20 fs, 0 fs]) and
  --|      (X = '1' during [-18 fs, 0 fs]) then
  --|   finally (Z = '1' during[0 fs, 14 fs]);
  --| elsewhen (Y = '0' during [-15 fs, 0 fs]) or
  --|          (X = '0' during [-14 fs, 0 fs]) then
  --|   finally (Z = '0' during[0 fs, 18 fs]);
  --| else
  --|   finally (Z = Z[-1 fs]);
  --| end when;
```

```
end CRAZY_AND;
```

No assertion violations are produced when this specification is transformed and simulated.

This specification is more complicated than the VHDL description. However, it makes explicit the timing behavior implicit in the VHDL description. It is exactly the implicit nature of the VHDL timing constructs that got us into trouble in the first place.

5.3.2 Altering the Implementation

Fortunately there is a fairly simple way in VHDL to produce an architecture that satisfies the original VAL specification. This can be accomplished using transport delay instead of inertial delay. Simply put, transport delay signal assignments only preempt events scheduled for a time greater than them in the event queue (Chapter 3).

The new architecture becomes:

```
-- VHDL architecture
architecture A of CRAZY_AND is
    signal X1, Y1 : bit;
begin
    X1 <= transport '1' after 18 fs when X = '1' else
                    '0' after 14 fs;
    Y1 <= transport '1' after 20 fs when Y = '1' else
                    '0' after 15 fs;
    Z <= X1 and Y1;
end A;
```

Simulating the above architecture with the original specification produces no assertion errors. (Fortunately, VAL is one of those tools where no news is good news, so if everything works correctly the simulation output is uninteresting and therefore we have not reproduced it here.)

Chapter 6

D-Type Flip-flop

6.1 Requirements

We wish to specify the entity behavior of a D-type flip-flop with timing parameters. The flip-flop under consideration is generic in three timing parameters, the set-up time (**SETUP**), the hold time (**HOLD**), and the propagation delay (**DELAY**). It has two inputs, **D** (data) and **Clk** (the triggering signal or clock); and two outputs, the state (**Q**) and its complement (**Qbar**). At each falling edge of the clock, the flip-flop updates its state to the input value, with a delay specified by the **DELAY** generic parameter. For the flip-flop to work properly, the input data must be stable from time **SETUP** before the calling edge of **Clk** to time **HOLD** (for example 5ns) after the falling edge of **Clk**.

The flip-flop is described in more detail in Chapter 2. Figure 2.7 in that chapter is a timing diagram showing the relationship between the various timing parameters.

6.2 Entity Declaration

The entity declaration is taken from Figure 2.2 in Chapter 2.

```
entity DFF is
  generic (SETUP, HOLD, DELAY : TIME);
  port (Clk : in BIT;  -- Clock input
```

```
        D    : in BIT;   -- Data input
        Q    : out BIT;  -- Output
        Qbar : out BIT); -- and its complement
   --| assume DELAY >= HOLD
   --|   report "Error in generic constant";
   --| state model is BIT;    -- A single bit of memory
begin
   --| when Clk'Changed('0') then
   --|   when D'Stable during [-SETUP, HOLD] then
   --|     D -> State[DELAY];
   --|   else
   --|     assert false report "SETUP-HOLD violation" ;
   --|   end when;
   --| end when;
   --| finally (State = Q) and ((not State) = Qbar)
   --|      report "State does not agree with output";
end DFF;
```

6.3 Commentary

The D-Type flip-flop example has a VAL entity state of predefined type BIT. The timing behavior of the D-Type flip-flop is specified using timed expressions and time qualified expressions.

Notice how the nested guards allow us to structure the annotation very much like the informal description of the first section. The outermost guard (**Clk'Changed ('0')**) captures the clock behavior. That behavior however is dependent on the stability of the input around the clock falling edge. That condition is checked by the innermost guard.

Notice also the separation of the annotation into two parts, corresponding to the two top level statements. The first statement expresses the flip-flop state maintenance. The second statement is an assertion about the outputs. This is very typical of the VAL annotation methodology. One part of the annotation updates the entity state using the current value of the state and the events on the entity inputs. The other part of the annotation expresses relations between the state value and the outputs, in the form of assertions. Those assertions need not be checked at every delta: flavors of assertions (in this case, the **finally** flavor) and guards give the user control of the exact simulation cycle where assertions are checked.

Finally, notice the assumption following the port declaration. The assumption expresses the designer's intention about the generic parameters. It does not make any sense to have a device whose propagation delay (`delay`) is smaller than the hold time (`hold`). This fact may not be obvious from the VAL statements. It is therefore a good idea to express such a constraint explicitly.

Consider now a VHDL architecture for a D-type flip-flop:

```
architecture a of dff is
  signal value : bit;
begin
  process(d,clk)
  begin
    if ((not clk'stable) and (clk = '0')) then
      value <= d;
    end if;
  end process;
  q <= transport value after delay;
  qbar <= transport not value after delay;
end a;
```

We have neglected the setup and hold times in the architecture on purpose. Often such delays are a result of the physical implementation, and really make no sense in a behavioral description. However, if the setup and hold times are violated in the simulation, then the actual device may not function properly when built. Here we rely on VAL to check that the device is used properly, and leave the setup and hold constraints out of the VHDL architecture.

We will use the following test bench and configuration to simulate the flip-flop:

```
entity tb is
end tb;

architecture a of tb is
  component flipflop_c
    generic(setup, hold, delay : time);
    port(clk,d : in bit;
         q,qbar : out bit);
  end component;
```

```
  signal s1,s2,s3,s4 : bit;
begin
  flipflop1 : flipflop_c
      generic map(2 ns, 1 ns, 5 ns)
      port map(s1,s2,s3,s4);

  s1 <= '1' after 5 ns,
        '0' after 15 ns,
        '1' after 25 ns,
        '0' after 30 ns,
        '1' after 35 ns,
        '0' after 40 ns;
  s2 <= '1' after 13 ns,
        '0' after 17 ns,
        '1' after 27 ns,
        '0' after 35 ns;
end a;

use work.all;
configuration c of tb is
  for a
    for flipflop1 : flipflop_c
      use entity dff(a);
      --| valentity;
    end for;
  end for;
end c;
```

Transforming and simulating the D-type flip-flop produces the following simulation results:

```
 TIME    |--------------------SIGNAL NAMES-------------------|
         |
 (FS)    |        CLK         D          Q          Qbar
         |
       0 |        '0'        '0'        '0'        '0'
      +1 |                              '0'        '0'
15000000 |        '1'
      +1 |                                         '1'
13000000 |                   '1'
15000000 |        '0'
16000000 | %VHDSIM-W-ASSERTV Assertion Violation after 16 ns
         |     at line 113 in design unit <<LMA.BOOK.DFF>>VALOU
         | T_DFF(STATEMONITOR)
         | setup-hold violation
17000000 |                   '0'
```

```
20000000 |
      +1 |                                  '1'         '0'
20000001 | %VHDSIM-W-ASSERTV Assertion Violation after 20000001
         |  fs
         |     at line 131 in design unit <<LMA.BOOK.DFF>>VALOU
         | T_DFF(STATEMONITOR)
         | state does not agree with output
25000000 |        '1'
27000000 |                   '1'
30000000 |        '0'
35000000 |        '1'        '0'
40000000 |        '0'
45000000 |
      +1 |                                  '0'         '1'
```

The assertion violation at 16 ns is due to the clock changing to `'0'` at 15 ns and the data changing to `'1'` at 13 ns. The setup and hold constraint requires the data to be stable for 2 ns before the clock edge and 1 ns after. The bounds are inclusive; ie. the data must be stable during time points 13 ns, 14 ns, 15 ns, and 16 ns. The architecture should be altered to include the setup and hold times described in the VAL specification.

The violation is not reported until 16 ns for a very subtle reason. The condition being checked is that the data is stable in the interval `[-2 ns, 1 ns]` about the clock change. The VAL Transformer implements this as a single check at the 1 ns point in the interval. Thus the violation is not detected until the expression over the interval can be completely evaluated (at the 16 ns).

Chapter 7

Traffic Light Controller

7.1 Requirements

This example of a traffic light controller is from "Introduction to VLSI" design by Mead and Conway [19] (pages 85-88). We reproduce here the informal specification as stated in the reference.

> A busy highway is intersected by a little used farmroad. A detector is installed on the farmroad, and reports the presence of cars wishing to cross or turn on the highway. There are lights controlling the traffic on the highway and on the farmroad. We wish to control the traffic lights so that, in the absence of any car on the farmroad, the highway lights will remain green, and the farmroad lights will remain red. If any car is detected on the farmroad, we wish the highway lights to cycle to yellow to red, and the farmroad lights to then turn to green. The farmroad lights are to remain green only while the detector senses car on the farmroad, but never longer than a fraction of a minute. The farmroad lights are then to cycle through yellow to red, and the highway lights then to turn to green. The highway lights are not to be interruptible again by the farmroad traffic until some fraction of a minute has passed.

7.2 Entity Declaration

The construction of the VAL specification for the traffic light controller is discussed in detail in Chapter 4. We briefly highlight here some of the important points.

```
entity dff is
  port(D, CK : in bit;
       Q     : out bit);
end dff;

architecture a of dff is
  signal value : bit;
begin
  process(d,ck)
  begin
    if ((not ck'stable) and (ck = '0')) then
      value <= d;
    end if;
  end process;
  q <= transport value after 5 ns;
end a;

package TLC_TYPES is
  type COLOR is (RED, YELLOW, GREEN);
end TLC_TYPES;

use work.TLC_TYPES.all;
entity TLC is
  generic (LTO, STO : TIME);
  port (CAR_ON_FARMRD : in boolean;
        HW_LIGHT, FM_LIGHT : out COLOR);
  --| constant SLOP : TIME := 50 ns;
begin
--| when (CAR_ON_FARMRD and
--|         (HW_LIGHT = GREEN) during [-LTO, 0 ns]) then
--|   finally ((HW_LIGHT'changed(YELLOW)) within [0 ns, SLOP])
--|     else report "Highway light did not change from " &
--|                 "green to yellow on time";
--| end when;

--| when (HW_LIGHT'changed(YELLOW)) then
--|   finally ((CAR_ON_FARMRD and ((HW_LIGHT = GREEN)
--|          during [-LTO, 0 ns])) within [-SLOP, 0 ns])
```

```
--|      else report "Highway light changed to yellow at " &
--|                  "incorrect time";
--| end when;

--| when ((HW_LIGHT = YELLOW) during [-STO, 0 ns]) then
--|   finally ((HW_LIGHT'changed(RED)) within [0 ns, SLOP])
--|     else report "Highway light did not change from " &
--|                 "yellow to red on time";
--|   finally ((FM_LIGHT'changed(GREEN)) within [0 ns, SLOP])
--|     else report "Farm road light did not change to " &
--|                 "green on time";
--| end when;

--| when (HW_LIGHT'changed(RED)) then
--|   finally (((HW_LIGHT = YELLOW) during [-STO, 0 ns])
--|          within [-SLOP, 0 ns])
--|     else report "Highway light changed to red at " &
--|                 "incorrect time";
--| end when;
--| when (FM_LIGHT'changed(GREEN)) then
--|   finally (((HW_LIGHT = YELLOW) during [-STO, 0 ns])
--|          within [-SLOP, 0 ns])
--|     else report "Farm road light changed to green at " &
--|                 "incorrect time";
--| end when;

--| when ((FM_LIGHT = GREEN) and (not CAR_ON_FARMRD)) or
--|       ((FM_LIGHT = GREEN) during [-LTO, 0 ns]) then
--|  finally ((FM_LIGHT'changed(YELLOW)) within [0 ns, SLOP])
--|     else report "Farm road light did not change to " &
--|                 "yellow on time";
--| end when;

--| when FM_LIGHT'changed(YELLOW) then
--|   finally (((FM_LIGHT = GREEN and not CAR_ON_FARMRD) or
--|           ((FM_LIGHT = GREEN) during [-LTO, 0 ns]))
--|           within [-SLOP, 0 ns])
--|     else report "Farm road light changed to yellow at " &
--|                 "incorrect time";
--| end when;

--| when ((FM_LIGHT = YELLOW) during [-STO, 0 ns]) then
--|   finally ((FM_LIGHT'changed(RED)) within [0 ns, SLOP])
--|     else report "Farm road light did not change from " &
--|                 "yellow to red on time";
--|   finally ((HW_LIGHT'changed(GREEN)) within [0 ns, SLOP])
```

```
--|      else report "Highway light did not change to " &
--|                  "green on time";
--| end when;

--| when FM_LIGHT'changed(RED) then
--|   finally (((FM_LIGHT = YELLOW) during[-STO, 0 ns])
--|           within [-SLOP, 0 ns])
--|      else report "Farm road light changed to red at " &
--|                  "incorrect time";
--| end when;
--| when HW_LIGHT'changed(GREEN) then
--|   finally (((FM_LIGHT = YELLOW) during[-STO, 0 ns])
--|           within [-SLOP, 0 ns])
--|      else report "Highway light changed to green at " &
--|                  "incorrect time";
--| end when;

--| assert not ((FM_LIGHT = GREEN) and
--|             (HW_LIGHT = GREEN)) else
--|      report "Danger - Farm road and highway lights " &
--|             "are both green";
end TLC;
```

The specification depends on two timing delays, STO and LTO. The short timeout (STO) determines the length of a yellow light. The long time out (LTO) determines the minimum time that the highway light must remain green, and the maximum time that the farm road light may remain green.

The input to the controller is a boolean signal from a detector that indicates the presence of a car on the farm road. The outputs of the controller are meant to control the two sets of traffic lights (highway and farm road).

Given the current state of the lights and inputs, the specification constrains when the lights may change. The specification is "loose" in that it allows some delay in the response of the implementation. The amount of delay in the response is determined by the timing constant SLOP.

7.3 Architecture

The architecture is a mixed VHDL data-flow and structural style description of the TLC. A set of 4 D-Type flip-flops are used to represent the state. The timers are implemented using VHDL delayed conditional signal assignment statements.

```
architecture A of TLC is
    constant CYCLE_TIME : TIME := 100 ns;
    signal QHG, DHG : bit := '1';
    signal QFG, DFG, QFY, DFY, QHY, DHY : bit := '0';
    signal ST, LT: bit := '0';
    signal CLK    : bit := '0';
    component dff_c
        port(D, CK : in bit;
             Q     : out bit);
    end component;
begin
    SHY : dff_c port map (DHY, CLK, QHY);
    SFG : dff_c port map (DFG, CLK, QFG);
    SFY : dff_c port map (DFY, CLK, QFY);
    SHG : dff_c port map (DHG, CLK, QHG);
    CLK <= not CLK after (CYCLE_TIME / 2);
    DHY <= '1' when ((QHG = '1') and (LT = '1') and
                      (CAR_ON_FARMRD)) or
                     ((QHY = '1') and not (ST = '1'))
               else '0';
    DHG <= '1' when ((QFY = '1') and (ST = '1')) or
                     ((QHG = '1') and not (CAR_ON_FARMRD and
                                            LT = '1')) or
                     ((QHY = '0') and (QFG = '0') and
                      (QFY = '0') and (QHG = '0'))
               else '0';
    DFY <= '1' when ((QFG = '1') and
                     ((LT = '1') or (not CAR_ON_FARMRD))) or
                     ((QFY = '1') and not (ST = '1'))
               else '0';
    DFG <= '1' when ((QHY = '1') and (ST = '1')) or
                     ((QFG = '1') and CAR_ON_FARMRD and
                      (LT = '0'))
               else '0';
    LT <= '1' after LTO when ((QFG = '1') or (QHG = '1'))
          else '0' after 1 fs;
    ST <= '1' after STO when ((QFY = '1') or (QHY = '1'))
          else '0' after 1fs;
```

```
    HW_LIGHT <= GREEN when  QHG = '1' else
                YELLOW when  QHY = '1' else
                RED;
    FM_LIGHT <= GREEN when  QFG = '1' else
                YELLOW when  QFY = '1' else
                RED;
--| assert (((QHY = '1') and (QFG = '0') and
--|           (QFY = '0') and (QHG = '0')) or
--|          ((QHY = '0') and (QFG = '1') and
--|           (QFY = '0') and (QHG = '0')) or
--|          ((QHY = '0') and (QFG = '0') and
--|           (QFY = '1') and (QHG = '0')) or
--|          ((QHY = '0') and (QFG = '0') and
--|           (QFY = '0') and (QHG = '1')))
--|   else report "Conflicting value for state";

end A;

entity tb is end tb;

use work.TLC_TYPES.all;
use std.simulator_standard;
architecture a of tb is
  component testentity
    generic (LTO, STO : TIME);
    port (CAR_ON_FARMRD : in boolean;
          HW_LIGHT, FM_LIGHT : out COLOR);
  end component;
  signal car_on_farmrd : boolean;
  signal hw_light, fm_light : color;
  signal terminate : boolean := false;
begin
  C1 : testentity
    generic map(700 ns, 350 ns)
    port map(car_on_farmrd, hw_light, fm_light);
  car_on_farmrd <= true after 450 ns,
                   false after 2850 ns,
                   true after 3300 ns,
                   false after 4100 ns;
  terminate <= true after 5300 ns;

  process (terminate)
  begin
    if (terminate) then
      simulator_standard.terminate;
    end if;
```

```
    end process;
end a;

use work.all;
configuration c of tb is
  for a
    for C1 : testentity
      use entity tlc(a);
      --| valentity;
      --| valarchitecture;
      for a
        for all : dff_c
          use entity dff(a);
        end for;
      end for;
    end for;
  end for;
end c;
```

7.4 Simulation Results

```
 TIME     |-------------------SIGNAL NAMES-------------------|
          |
 (FS)     |    F      H      C     D    D    D    D    L    S    C
          |    M      W      A     F    F    H    H    T    T    L
          |    _      _      R     G    Y    G    Y              K
          |    L      L      _
          |    I      I      O
          |    G      G      N
          |    H      H      _
          |    T      T      F
          |                  A
          |                  R
          |                  M
          |                  R
          |                  D
          |
        0 | %VHDSIM-W-ASSERTV Assertion Violation after 0 fs
          |     at line 1077 in design unit <<LMA.BOOK.TLC>>VA
          | LIN_TLC(VAL_A)
          | conflicting value for state
          |   RED    RED    FALSE '0' '0' '1' '0' '0' '0' '0'
       +1 |   RED    RED          '0' '0' '1' '0'
        1 |                                         '0' '0'
```

```
   50000000 |                                                  '1'
  100000000 |                                                  '0'
  105000000 |
         +1 |                                '1' '0'
         +3 |        GREEN
  105000001 | %VHDSIM-W-ASSERTV Assertion Violation after 105000
            | 001 fs
            |     at line 935 in design unit <<LMA.BOOK.TLC>>VAL
            | OUT_TLC(STATEMONITOR)
            | highway light changed to green at incorrect time
  150000000 |                                                  '1'
  200000000 |                                                  '0'
  250000000 |                                                  '1'
  300000000 |                                                  '0'
  350000000 |                                                  '1'
  400000000 |                                                  '0'
  450000000 |               TRUE                               '1'
         +1 |                      '0' '0' '1' '0'
  500000000 |                                                  '0'
  550000000 |                                                  '1'
  600000000 |                                                  '0'
  650000000 |                                                  '1'
  700000000 |                                                  '0'
  750000000 |                                                  '1'
  800000000 |                                                  '0'
  805000000 |                                      '1'
         +1 |                      '0' '0' '0' '1'
  850000000 |                                                  '1'
  900000000 |                                                  '0'
  905000000 |
         +1 |                      '0'     '0' '1'
         +3 |        YELLOW
  905000001 |                                      '0'
         +1 |                      '0' '0' '0' '1'
  950000000 |                                                  '1'
 1000000000 |                                                  '0'
 1050000000 |                                                  '1'
 1100000000 |                                                  '0'
 1150000000 |                                                  '1'
 1200000000 |                                                  '0'
 1250000000 |                                                  '1'
 1255000000 |                                          '1'
         +1 |                      '1' '0' '0' '0'
 1300000000 |                                                  '0'
 1305000000 |
         +1 |                      '1' '0' '0' '0'
         +3 |  GREEN   RED
 1305000001 |                                          '0'
```

```
        +1 |                    '1' '0' '0' '0'
1350000000 |                                            '1'
1400000000 |                                            '0'
1450000000 |                                            '1'
1500000000 |                                            '0'
1550000000 |                                            '1'
1600000000 |                                            '0'
1650000000 |                                            '1'
1700000000 |                                            '0'
1750000000 |                                            '1'
1800000000 |                                            '0'
1850000000 |                                            '1'
1900000000 |                                            '0'
1950000000 |                                            '1'
2000000000 |                                            '0'
2005000000 |                                      '1'
        +1 |                    '0' '1' '0' '0'
2050000000 |                                            '1'
2055000001 | %VHDSIM-W-ASSERTV Assertion Violation after 205500
           | 0001 fs
           |    at line 782 in design unit <<LMA.BOOK.TLC>>VAL
           | OUT_TLC(STATEMONITOR)
           | farm road light did not change to yellow on time
2100000000 |                                            '0'
2105000000 |
        +1 |                    '0' '1' '0'
        +3 | YELLOW
2105000001 |                                      '0'
        +1 |                    '0' '1' '0' '0'
2150000000 |                                            '1'
```

The simulation has produced 3 assertion violations. The first violation occurs because the initial values of the state signals in the architecture are `'0'` in VHDL. This is not one of the legal states as specified by the assertion appearing in the architecture. The second violation also occurs because of a problem with initialization. The architecture moves into a legal state with the highway light green and the farm road light red at the first clock cycle after the simulation begins. The VAL specification requires that the highway light change to green only when the farm road light has been yellow for the short time out period. Since the farm road light has been red from the beginning of the simulation, this is clearly a violation.

Neither of these violations is particularly serious. We could correct them by several methods. The specification might be changed to allow the red–red state. The implementation might be changed to come up in a different initial state. A reset signal might be added to the device, and the specification modified to only apply after the device has been reset and is in a running state. We leave the modifications as an exercise to the reader.

The third violation is more significant. The highway light changed to yellow 800 ns after it changed to green. According to the specification, the farm road light can stay green for at most 350 ns (STO) + 50 ns (SLOP). The reason for the delay in the change can be traced to the clocked nature of the implementation. Examining the architecture, we see that if the short time out timer expires just after a falling clock edge, the device will not react to this change until the next falling clock edge plus the delay through the logic following the clock. Since a clock cycle is 100 ns, but the amount of slack allowed by the specification is only 50 ns, a violation results. The violation may be corrected by improving the implementation so that it responds faster, or by modifying the specification. If the customer is happy with a slightly longer delay in response, then the amount of slop may be increased. Some compromise of speeding up the clock and relaxing the constraint is also an alternative.

If we modify the SLOP parameter making it 110 ns and simulate, the following trace is obtained. Note that the violation at 2055 ns no longer occurs.

```
    TIME    |------------------SIGNAL NAMES------------------|
            |
    (FS)    |    F      H      C    D    D    D    D    L    S    C
            |    M      W      A    F    F    H    H    T    T    L
            |    _      _      R    G    Y    G    Y              K
            |    L      L      _
            |    I      I      O
            |    G      G      N
            |    H      H      _
            |    T      T      F
            |                  A
            |                  R
            |                  M
            |                  R
            |                  D
            |
          0 | %VHDSIM-W-ASSERTV Assertion Violation after 0 fs
            |     at line 1077 in design unit <<LMA.BOOK.TLC>>VA
```

```
           | LIN_TLC(VAL_A)
           | conflicting value for state
           |   RED    RED   FALSE '0' '0' '1' '0' '0' '0' '0'
        +1 |   RED    RED         '0' '0' '1' '0'
         1 |                                      '0' '0'
  50000000 |                                              '1'
 100000000 |                                              '0'
 105000000 |
        +1 |                              '1' '0'
        +3 |         GREEN
 105000001 | %VHDSIM-W-ASSERTV Assertion Violation after 105000
           | 001 fs
           |     at line 935 in design unit <<LMA.BOOK.TLC>>VAL
           | OUT_TLC(STATEMONITOR)
           | highway light changed to green at incorrect time
 150000000 |                                              '1'
 200000000 |                                              '0'
 250000000 |                                              '1'
 300000000 |                                              '0'
 350000000 |                                              '1'
 400000000 |                                              '0'
 450000000 |                TRUE                          '1'
        +1 |                      '0' '0' '1' '0'
 500000000 |                                              '0'
 550000000 |                                              '1'
 600000000 |                                              '0'
 650000000 |                                              '1'
 700000000 |                                              '0'
 750000000 |                                              '1'
 800000000 |                                              '0'
 805000000 |                                      '1'
        +1 |                      '0' '0' '0' '1'
 850000000 |                                              '1'
 900000000 |                                              '0'
 905000000 |
        +1 |                      '0'     '0' '1'
        +3 |         YELLOW
 905000001 |                                      '0'
        +1 |                      '0' '0' '0' '1'
 950000000 |                                              '1'
1000000000 |                                              '0'
1050000000 |                                              '1'
1100000000 |                                              '0'
1150000000 |                                              '1'
1200000000 |                                              '0'
1250000000 |                                              '1'
1255000000 |                                          '1'
        +1 |                      '1' '0' '0' '0'
```

```
1300000000 |                                               '0'
1305000000 |
        +1 |                     '1' '0' '0' '0'
        +3 |  GREEN   RED
1305000001 |                                           '0'
        +1 |                     '1' '0' '0' '0'
1350000000 |                                               '1'
1400000000 |                                               '0'
1450000000 |                                               '1'
1500000000 |                                               '0'
1550000000 |                                               '1'
1600000000 |                                               '0'
1650000000 |                                               '1'
1700000000 |                                               '0'
1750000000 |                                               '1'
1800000000 |                                               '0'
1850000000 |                                               '1'
1900000000 |                                               '0'
1950000000 |                                               '1'
2000000000 |                                               '0'
2005000000 |                                       '1'
        +1 |                     '0' '1' '0' '0'
2050000000 |                                               '1'
2100000000 |                                               '0'
2105000000 |
        +1 |                     '0' '1' '0'
        +3 |  YELLOW
2105000001 |                                       '0'
        +1 |                     '0' '1' '0' '0'
2150000000 |                                               '1'
2200000000 |                                               '0'
2250000000 |                                               '1'
2300000000 |                                               '0'
2350000000 |                                               '1'
2400000000 |                                               '0'
2450000000 |                                               '1'
2455000000 |                                           '1'
        +1 |                     '0' '0' '1' '0'
2500000000 |                                               '0'
2505000000 |
        +1 |                         '0' '1' '0'
        +3 |   RED   GREEN
2505000001 |                                           '0'
        +1 |                     '0' '0' '1' '0'
2550000000 |                                               '1'
2600000000 |                                               '0'
2650000000 |                                               '1'
2700000000 |                                               '0'
2750000000 |                                               '1'
```

```
2800000000 |                                                '0'
2850000000 |                 FALSE                          '1'
        +1 |                       '0' '0' '1' '0'
2900000000 |                                                '0'
2950000000 |                                                '1'
3000000000 |                                                '0'
3050000000 |                                                '1'
3100000000 |                                                '0'
3150000000 |                                                '1'
3200000000 |                                                '0'
3205000000 |                                       '1'
        +1 |                       '0' '0' '1' '0'
3250000000 |                                                '1'
3300000000 |                 TRUE                           '0'
        +1 |                       '0' '0' '0' '1'
3350000000 |                                                '1'
3400000000 |                                                '0'
3405000000 |
        +1 |                       '0'     '0' '1'
        +3 |          YELLOW
3405000001 |                                       '0'
        +1 |                       '0' '0' '0' '1'
3450000000 |                                                '1'
3500000000 |                                                '0'
3550000000 |                                                '1'
3600000000 |                                                '0'
3650000000 |                                                '1'
3700000000 |                                                '0'
3750000000 |                                                '1'
3755000000 |                                           '1'
        +1 |                       '1' '0' '0' '0'
3800000000 |                                                '0'
3805000000 |
        +1 |                       '1' '0' '0' '0'
        +3 |  GREEN   RED
3805000001 |                                           '0'
        +1 |                       '1' '0' '0' '0'
3850000000 |                                                '1'
3900000000 |                                                '0'
3950000000 |                                                '1'
4000000000 |                                                '0'
4050000000 |                                                '1'
4100000000 |                 FALSE                          '0'
        +1 |                       '0' '1' '0' '0'
4150000000 |                                                '1'
4200000000 |                                                '0'
4205000000 |
        +1 |                       '0' '1' '0'
        +3 |  YELLOW
```

```
4205000001 |                                        '0'
4250000000 |                                                '1'
4300000000 |                                                '0'
4350000000 |                                                '1'
4400000000 |                                                '0'
4450000000 |                                                '1'
4500000000 |                                                '0'
4550000000 |                                                '1'
4555000000 |                                            '1'
        +1 |                       '0' '0' '1' '0'
4600000000 |                                                '0'
4605000000 |
        +1 |                           '0' '1' '0'
        +3 |    RED   GREEN
4605000001 |                                            '0'
        +1 |                       '0' '0' '1' '0'
4650000000 |                                                '1'
4700000000 |                                                '0'
4750000000 |                                                '1'
4800000000 |                                                '0'
4850000000 |                                                '1'
4900000000 |                                                '0'
4950000000 |                                                '1'
5000000000 |                                                '0'
5050000000 |                                                '1'
5100000000 |                                                '0'
5150000000 |                                                '1'
5200000000 |                                                '0'
5250000000 |                                                '1'
5300000000 |                                                '0'
```

Chapter 8

Stack

8.1 Requirements

This example describes the verification of the hardware stack described in VAL/VHDL in Section 4.3. The hardware stack stores elements of type **ElementT** and has a maximum depth **Depth**. The specification and implementation are both written so as to be independent of the particular choice of **Depth** or **ElementT**. The functions abstracting the computation of the stack are described in a package **StackP**. This package defines an internal data type representing the stack, as well as a set of functions (**SPush**, **SPop**, etc.) operating on the stack. The relationship between the functions (constraints on when the computation occurs) is given in the entity declaration. A set of VAL annotations constrain the order of computation.

8.2 Entity Declaration

The VHDL entity declaration of the stack specifies the I/O ports of the device. Our stack has separate ports for the input and output of data, a direction indicating whether a push or pop is to be performed, a clock input whose rising edge actually causes the push or pop, and two status output lines that indicate if the stack is full or empty.

The VAL annotations specify constraints on the behavior of the stack. The first **when** clause maintains the internal state of the stack. When there is a rising edge on the clock the stack performs a push or pop depending on the direction signal. The remaining assertions check that the other output pins are consistent with the internal state of the stack.

In order to separate the constraints on the stack from its implementation, we construct a stack package **StackP** for modeling the state of the stack. The annotations appearing in the entity declaration should be abstract, referring only to operations such as **SPush**, **SPop**, etc. provided by the package. Unfortunately, VHDL does not support information hiding in packages through the use of private types as does Ada, so the definition of the stack data type **StackT** is visible. In addition, VHDL does not support generic packages, so the stack package cannot be generic in the type of the elements it contains. Instead, we define the element type in a separate package, and import that package into the stack package.

```
package Element is
    subtype ElementT is bit_vector(0 to 3);
    constant NullElement   : ElementT := ('0','0','0','0');
    constant OneElement    : ElementT := ('0','0','0','1');
    constant TwoElement    : ElementT := ('0','0','1','0');
    constant ThreeElement  : ElementT := ('0','0','1','1');
    constant FourElement   : ElementT := ('0','1','0','0');
    constant FiveElement   : ElementT := ('0','1','0','1');
    constant SixElement    : ElementT := ('0','1','1','0');
end Element;

use work.Element.all;
package StackP is
    constant depth : natural := 5;
    type stackarray is array(natural range 1 to depth) of
            ElementT;
    type StackT is record
            top : natural;
            stack : stackarray;
    end record;
    constant EmptyStack : StackT :=
            StackT'(0,StackArray'(others=>NullElement));
    function SPush(s : in StackT; e : ElementT)
            return StackT;
    function SPop(s : in StackT) return StackT;
    function STop(s : StackT) return ElementT;
    function is_Full(s : StackT) return boolean;
    function is_Empty(s : StackT) return boolean;
end StackP;
```

```
package body stackp is
    function SPush(s : in StackT; e : ElementT)
            return StackT is
        variable t : StackT;
    begin
        if (s.top < depth) then
            t := s;
            t.top := t.top + 1;
            t.stack(t.top) := e;
            return t;
        else
            return s;
        end if;
    end SPush;

    function SPop(s : in StackT) return StackT is
        variable t : StackT;
    begin
        if (s.top > 0) then
            t := s;
            t.top := t.top - 1;
            return t;
        else
            return s;
        end if;
    end SPop;

    function STop(s : StackT) return ElementT is
    begin
        if (s.top > 0) and (s.top <= depth) then
            return s.stack(s.top);
        else
            return NullElement;
        end if;
    end STop;

    function is_Full(s : StackT) return boolean is
    begin
        if (s.top = depth) then
            return true;
        else
            return false;
        end if;
    end is_Full;

    function is_Empty(s : StackT) return boolean is
```

```
    begin
        if (s.top = 0) then
            return true;
        else
            return false;
        end if;
    end is_Empty;
end stackp;

use work.Element.all;
--| use work.StackP.all;
entity Stack is
    port(din   : in ElementT;      -- data input
         dout  : out ElementT;     -- data output
         dir   : in bit;           -- push or pop
         clk   : in bit;           -- do the operation
         ce    : in bit;           -- chip enable
         full  : out bit;          -- is the stack full?
         empty : out bit);         -- is the stack empty?
    --| state model is StackT := EmptyStack;
begin
--| when (clk'changed('1') and (ce = '1')) then
--|     when (dir = '1') then
--|         when not (is_full(state)) then
--|             state <- SPush(state,din);
--|         else
--|             assert false
--|                 report "push attempted on full stack";
--|         end when;
--|     else
--|         when (not is_empty(state)) then
--|             state <- SPop(state);
--|         else
--|             assert false
--|                 report "pop attempted on empty stack";
--|         end when;
--|     end when;
--| end when;
--| when(not is_empty(state)) then
--|     finally (dout = STop(state))
--|         report "top of stack and state do not agree";
--| end when;
--| when (is_empty(state)) then
--|     finally(empty = '1')
--|         report "empty signal should be 1";
--| else
--|     finally(empty = '0')
```

```
--|          report "empty signal should be 0";
--| end when;
--| when (is_full(state)) then
--|     finally(full = '1')
--|          report "full signal should be 1";
--| else
--|     finally(full = '0')
--|          report "full signal should be 0";
--| end when;
end Stack;
```

Note that the entity specification avoids describing "how" the operations on the stack are implemented. Instead, it just constrains the current output of the device (`dout`) along with the `empty` and `full` signals. The constraints use operations of package `StackP`. We note that these specifications do not express "stack" semantics. If we wish to apply formal proof methods to the VAL stack, we would need the `StackP` package specification to contain an abstract algebraic specification of its operations. Stack semantics would then be imposed on the entity stack because of its use of the operations of `StackP`.

In this example, the operations `SPush`, `SPop`, and `STop` are implemented as stack operations in the body of package `StackP`.

8.3 Entity Architecture

Here we present the stack architecture shown in Figure 4.4. An array of registers are connected by multiplexors. For the registers, we also have a VAL specification of the behavior of the register. Notice the state model annotation in the component declaration for `creg` in the stack architecture.

In addition, the architecture contains some more detailed annotations as described previously in Section 4.3. These annotations serve to add extra internal consistency checking inside the architecture. Notice in the configuration declaration that `valentity` and `valarchitecture` have been specified for the stack, indicating that we wish to transform and simulate using both the annotations in the entity declaration and those in the architecture. In addition, since the register also contains annotations as part of its entity declaration, `valentity` has been specified for it in the configuration.

```
architecture structure of Stack is
    type ElementVector is array(Natural range 1 to depth) of
                                                   ElementT;

    signal zero : bit := '0';
    signal one : bit := '1';

    component creg
        port(din : in ElementT;
             dout : out ElementT;
             en : in bit;
             ck : in bit);
        --| state model is ElementT := NullElement;
    end component;

    component cmux
        port(d1, d2 : in ElementT;
             dout : out ElementT;
             s : in bit);
    end component;

    component shiftreg
        generic(size : natural);
        port(lin : in bit;
             rin : in bit;
             dout : out bit_vector(1 to size);
             shift : in bit;
             dir : in bit);
    end component;

    signal mux0in : ElementVector;
    signal mux1in : ElementVector;
    signal muxout : ElementVector;
    signal regout : ElementVector;
    signal srout : bit_vector(1 to depth);
    signal enable, sshift : bit;

begin
    G1: for i in 1 to depth generate
        M1 : cmux
            port map(mux0in(i), mux1in(i), muxout(i), dir);
        R1 : creg
            port map(muxout(i), regout(i), enable, clk);
        G2 : if (i >= 2) generate
            mux0in(i - 1) <= regout(i);
```

```
            mux1in(i) <= regout(i - 1);
        end generate;
    end generate;

    mux1in(1) <= din;
    mux0in(depth) <= NullElement;

    SR : shiftreg
        generic map(depth)
        port map(one, zero, srout, sshift, dir);

    empty <= not srout(1);
    full <= srout(depth);

    enable <= ce and ((dir and (not srout(depth))) or
              ((not dir) and srout(1)));
    sshift <= ce and (enable and clk);
    dout <= regout(1);

--| for i in 1 to depth generate
--|   when (i <= state.top) then
--|     finally (state.stack(state.top + 1 - i) =
--|              G1.R1(i).state)
--|         report "internal state inconsistent with val " &
--|                "state model";
--|   end when;
--| end generate;

--| for i in 1 to depth generate
--|   when (i <= state.top) then
--|     finally srout(i) = '1'
--|         report "srout state inconsistent with val " &
--|                "state model: 1 expected";
--|   else
--|     finally srout(i) = '0'
--|         report "srout state inconsistent with val " &
--|                "state model: 0 expected";
--|   end when;
--| end generate;
end structure;

use work.Element.all;
entity reg is
    port(din : in ElementT;
         dout : out ElementT;
         en : in bit;
         ck : in bit);
```

```
--| state model is ElementT := NullElement;
begin
--| when (ck'changed('1') and (en = '1') and
--|        (not din'event)) then
--|     state <- din;
--| end when;
--| finally (dout = state)
--|     report "register output does not agree with state";
end reg;

architecture A of reg is
    signal value : ElementT;
begin
    process(ck)
    begin
        if ((ck = '1') and (en = '1')) then
            value <= din;
        end if;
    end process;
    dout <= value;
end A;

use work.Element.all;
entity mux is
    port(d1, d2 : in ElementT;
         dout : out ElementT;
         s : in bit);
end mux;

architecture A of mux is
begin
    process(d1, d2, s)
    begin
        if (s = '0') then
            dout <= d1;
        else -- (s = '1')
            dout <= d2;
        end if;
    end process;
end A;

entity shiftregister is
    generic(size : natural);
    port(lin : in bit;
         rin : in bit;
         dout : out bit_vector(1 to size);
         shift : in bit;
```

```
            dir : in bit);
  end shiftregister;

  architecture A of shiftregister is
      signal contents : bit_vector(1 to size);
  begin
      process(shift)
      begin
          if(shift = '1') then
              if(dir = '1') then -- shift right
                  for i in contents'right downto
                              (contents'left + 1) loop
                      contents(i) <= contents(i - 1);
                  end loop;
                  contents(1) <= lin;
              else -- shift left
                  for i in contents'left to
                              (contents'right - 1) loop
                      contents(i) <= contents(i + 1);
                  end loop;
                  contents(size) <= rin;
              end if;
          end if;
      end process;
      dout(1 to size) <= contents(1 to size);
  end A;

  entity tb is end tb;

  use work.Element.all;
  architecture a of tb is
      component testentity
          port(din : in ElementT;
               dout : out ElementT;
               dir : in bit;
               clk : in bit;
               ce  : in bit;
               full : out bit;
               empty : out bit);
      end component;

      signal din, dout : ElementT;
      signal dir, clk, full, empty, ce : bit;
      signal terminate : bit := '1';
  begin
      C1 : testentity
          port map(din, dout, dir, clk, ce, full, empty);
```

```
        terminate <= '1', '0' after 240 ns;
        ce <= '1' after 5 ns;
        clk <= (terminate and (not clk)) after 10 ns;
        dir <= '1', '0' after 120 ns;
        din <= OneElement, TwoElement after 15 ns,
            ThreeElement after 35 ns,
            FourElement after 55 ns,
            FiveElement after 75 ns,
            SixElement after 95 ns;
end a;
use work.all;

configuration c of tb is
    for a
        for all : testentity
            use entity stack(structure);
            --| valentity;
            --| valarchitecture;
            for structure
                for G1
                    for all : creg
                        use entity reg(a);
                        --| valentity;
                    end for;
                    for all : cmux
                        use entity mux(a);
                    end for;
                end for;
                for all : shiftreg
                    use entity shiftregister(a);
                end for;
            end for;
        end for;
    end for;
end c;
```

8.4 Commentary

Transforming and simulating the above description produces the following result:

```
TIME    |--------------------SIGNAL NAMES-------------------|
        |
(FS)    |  CLK  DIN(0 TO 3)  DIR  DOUT(0 TO 3)  EMPTY  FULL
```

```
          |
        0 | '0'     "0000"     '0'     "0000"      '0'   '0'
       +1 |         "0001"     '1'     "0000"      '0'   '0'
       +3 |                                        '1'
 10000000 | '1'
       +6 |                            "0001"      '0'
 15000000 |         "0010"
 20000000 | '0'
 30000000 | '1'
       +6 |                            "0010"
 35000000 |         "0011"
 40000000 | '0'
 50000000 | '1'
       +6 |                            "0011"
 55000000 |         "0100"
 60000000 | '0'
 70000000 | '1'
       +6 |                            "0100"
 75000000 |         "0101"
 80000000 | '0'
 90000000 | '1'
       +6 |                            "0101"            '1'
 95000000 |         "0110"
100000000 | '0'
110000000 | %VHDSIM-W-ASSERTV Assertion Violation after 110 ns
          |     at line 174 in design unit <<LMA.BOOK.STACK>>VA
          | LOUT_STACK(STATEMONITOR)
          | push attempted on full stack
          | '1'
120000000 | '0'                '0'
130000000 | '1'
       +6 |                            "0100"            '0'
140000000 | '0'
150000000 | '1'
       +6 |                            "0011"
160000000 | '0'
170000000 | '1'
       +6 |                            "0010"
180000000 | '0'
190000000 | '1'
       +6 |                            "0001"
200000000 | '0'
210000000 | '1'
       +6 |                            "0000"      '1'
220000000 | '0'
230000000 | %VHDSIM-W-ASSERTV Assertion Violation after 230 ns
          |     at line 185 in design unit <<LMA.BOOK.STACK>>VA
          | LOUT_STACK(STATEMONITOR)
```

```
          | pop attempted on empty stack
          |  '1'
240000000 |  '0'
250000000 |  '0'
```

The first assertion violation occurs because the test bench for the stack attempts to push too many items on to the stack, causing an overflow. The second violation occurs for exactly the opposite reason; too many items are popped off the stack causing an underflow.

While we have not guaranteed the correctness of the stack implementation, the fact that the simulation of the architecture agreed with the VAL specification on this particular test data increases our confidence in the correctness of the design.

Chapter 9

Water Heater Controller

9.1 Requirements

The water heater controller example is taken from the problem set of the Fourth International Workshop on Software Specification and Design, April 3-4, 1987, Monterey, California:

> Problem #2. Heating System. (Based on a problem by S. White presented to 1984 Embedded Computer System Requirement Workshop.)
>
> The controller of an oil hot water home heating system regulates in-flow of heat, by turning the furnace on and off, and monitors the status of combustion and fuel flow of the furnace system, provided the master switch is set to HEAT position. The controller activates the furnace whenever the home temperature, t, falls below $t_r - 2$ degrees, where t_r is the desired temperature set bt the user. The activation procedure is as follows:
>
> 1. The controller signals the motor to be activated.
> 2. The controller monitors the motor speed and once the speed is adequate it signals the ignition and oil valve to be activated.
> 3. The controller monitors the water temperature and once the temperature is reached [sic] a predefined value it

signals the circulation valve to be opened. The heated water then starts to circulate through the house.

4. A fuel flow indicator and an optical compbustion sensor signal the controller if abnormalities occur. In this case the controller signals the system to be shut off.
5. Once the home temperature reaches $t_r + 2$ degrees, the controller deactivates the furnace by first closing the oil valve and then, after 5 seconds, stopping the motor.

In addition, the system is subject to the fllowing constraints:

1. Minimum time for furnace restart after prior operation is 5 minutes.
2. Furnace turn off shall be indicated within 5 seconds if master switch shut off or fuel flow shut off.

9.2 Entity Declaration

The informal specification closely resembles a finite state machine, and we have used this style to construct the VAL specification. The entity state **HeatingSystem_state** model is an enumeration type consisting of seven abstract states, corresponding to the seven states of the informal specification.

In each state, we specify the allowed values of the outputs. Macros prove useful in constructing the output annotations. A parameterized macro (**AssertOutputs**) is used to succinctly describe the assertions on the output ports.

Timing is included using the **during** construct. The **during** expressions read just as the informal English specification.

```
package HeatingSystemType is
   type HeaterSwitchMode is (off, heat);
   type FlowMode is (normal, stop);
   type CombustionMode is (normal, danger);
   type SwitchMode is (off, onn);
   type ValveMode is (closed, openn);
   type HeatingSystemState is
```

```
        (Idle, MotorStart, Heating, Circulating,
         Waiting, DeActivate);
    subtype temperature is integer range 0 to 999;
    subtype speed is integer range 0 to 9999;
end HeatingSystemType;

use work.HeatingSystemType.all;
entity HeatingSystem is
   generic(HomeTempThreshold: temperature;
           WaterTempThreshold: temperature;
           MotorSpeedThreshold: speed;
           TurnOffDelay: time;
           RestartDelay: time);
   port(MasterSwitch: in HeaterSwitchMode;
        HomeTemp: in temperature;
        TempSetting: in temperature;
        MotorSpeed: in speed;
        WaterTemp: in temperature;
        FuelFlow: in FlowMode;
        Combustion: in CombustionMode;
        MotorSwitch: out SwitchMode;
        IgnitionSwitch: out SwitchMode;
        OilValve: out ValveMode;
        CirculationValve: out ValveMode);

--| state model is HeatingSystemState := Idle;
--| macro AssertOutputs(SName, MSwitch, ISwitch,
--|                     OValve, CValve, ErrMsg) is
--|   when (state = SName) then
--|     finally((MSwitch = MotorSwitch) and
--|             (ISwitch = IgnitionSwitch) and
--|             (OValve  = OilValve) and
--|             (CValve = CirculationValve))
--|     report ErrMsg;
--|   end when;
--| end AssertOutputs;

begin
--| when (MasterSwitch = off) or (FuelFLow = stop) or
--|      (Combustion = danger) then
--|   state <- DeActivate;
--| else
--|   when (state = Idle) and
--|        (HomeTemp < TempSetting - HomeTempThreshold) then
--|     state <- MotorStart;
--|   end when;
--|   when (state = MotorStart) and
```

```
--|         (MotorSpeed >= MotorSpeedThreshold) then
--|      state <- Heating;
--|    end when;
--|    when (state = Heating) and
--|         (WaterTemp >= WaterTempThreshold) then
--|      state <- Circulating;
--|    end when;
--|    when (state = Circulating) and
--|        (HomeTemp >= TempSetting + HomeTempThreshold) then
--|      state <- DeActivate;
--|    end when;
--|    when (state = DeActivate) during TurnOffDelay then
--|      state <- Waiting;
--|    end when;
--|    when (state = Waiting) during RestartDelay then
--|      state <- Idle;
--|    end when;
--|end when;

--| AssertOutputs(Idle, off, off, closed, closed,
--|               "Incorrect output in state idle");
--| AssertOutputs(MotorStart, onn, off, closed, closed,
--|               "Incorrect output in state motorstart");
--| AssertOutputs(Heating, onn, onn, openn, closed,
--|               "Incorrect output in state heating");
--| AssertOutputs(Circulating, onn, onn, openn, openn,
--|               "Incorrect output in state circulating");
--| AssertOutputs(DeActivate, MotorSwitch, off, closed,
--|               CirculationValve,
--|               "Incorrect output in state deactivate");
--| AssertOutputs(Waiting, off, off, closed, closed,
--|               "Incorrect output in state waiting");

end HeatingSystem;
```

Note how the annotation is divided into two parts; the state maintenance part, and the assertion part. The state maintenance is made of nested **when** statements. The outermost **when** separates the emergency behavior from the regular behavior. In case of emergency (master switch off, fuel flow stopped or combustion alarm), the water heater controller moves to state **shutdown**, and stays there as long as the emergency condition is true. Otherwise, the controller resumes normal behavior (**else** branch of the outermost **when** statement). The normal behavior is a number of parallel **when** statements, one for each state of the controller. The separation between emergency and normal behavior is very common in VAL descriptions.

Typically, VAL state machine specifications have the following structure:

```
when emergency_condition then
      reset_behavior
  else
      when state1 then
           ...
      end when;
      when state2 then
           ...
      end when;
      ...
  end when;
```

It is straightforward to specify both Mealy and Moore style state machines in this manner. The water heater controller is based on a Moore model. For readability, we chose to specify the outputs during a state using the **AssertOutputs** macro seperately from the state changes.

Note also how outputs whose value is a don't care in a particular state were specified using the **AssertOutputs** macro. In that case, the outputs were simply constrained to equal themselves.

9.3 Implementation

Since the specification of the water heater controller is already at a fairly low level (that of a state machine), constructing a VHDL implementation of it follows fairly directly. The implementation describes a state machine having the same states as the specification.

```
architecture a of HeatingSystem is
signal vhdl_state: HeatingSystemState := Idle;
begin
  process(MasterSwitch, HomeTemp, TempSetting, MotorSpeed,
          WaterTemp, FuelFlow, Combustion)
  begin
    if (MasterSwitch = off) or (FuelFLow = stop) or
```

```
          (Combustion = danger) then
        vhdl_state <= DeActivate;
        IgnitionSwitch <= off;
        OilValve <= closed;
      else
        if (vhdl_state = Idle) and
           (HomeTemp < TempSetting - HomeTempThreshold) then
          vhdl_state <= MotorStart;
          MotorSwitch <= onn;
        elsif (vhdl_state = MotorStart) and
              (MotorSpeed >= MotorSpeedThreshold) then
          vhdl_state <= Heating;
          IgnitionSwitch <= onn;
          OilValve <= openn;
        elsif (vhdl_state = Heating) and
              (WaterTemp >= WaterTempThreshold) then
          vhdl_state <= Circulating;
          CirculationValve <= openn;
        elsif (vhdl_state = Circulating) and
            (HomeTemp >= TempSetting + HomeTempThreshold) then
          vhdl_state <= DeActivate;
          IgnitionSwitch <= off;
          OilValve <= closed;
        elsif (vhdl_state = DeActivate) and
              vhdl_state'stable(TurnOffDelay) then
          vhdl_state <= Waiting;
          CirculationValve <= closed;
          MotorSwitch <= off;
        elsif (vhdl_state = Waiting) and
              vhdl_state'stable(RestartDelay) then
          vhdl_state <= Idle;
        end if;
      end if;
    end process;
  end a;

  use work.HeatingSystemType.all;
  entity tb is
  end tb;

  architecture a of tb is
    signal MaSw: HeaterSwitchMode := heat;
    signal HoTe: temperature := 60;
    signal TeSe: temperature :=  75;
    signal MoSp: speed := 0;
    signal WaTe: temperature := 60;
```

```
signal FuFl: FlowMode := normal;
signal Co   : CombustionMode :=  normal;
signal MoSw: SwitchMode := off;
signal IgSw: SwitchMode := off;
signal OiVa: ValveMode  := closed;
signal CiVa: ValveMode  := closed;

component whc_c
  generic(HomeTempThreshold: temperature;
          WaterTempThreshold: temperature;
          MotorSpeedThreshold: speed;
          TurnOffDelay: time;
          RestartDelay: time);
  port(MasterSwitch: in HeaterSwitchMode;
       HomeTemp: in temperature;
       TempSetting: in temperature;
       MotorSpeed: in speed;
       WaterTemp: in temperature;
       FuelFlow: in FlowMode;
       Combustion: in CombustionMode;
       MotorSwitch: out SwitchMode;
       IgnitionSwitch: out SwitchMode;
       OilValve: out ValveMode;
       CirculationValve: out ValveMode);
 end component;
begin
 whcl: whc_c
  generic map(HomeTempThreshold => 5,
              WaterTempThreshold => 75,
              MotorSpeedThreshold => 1000,
              TurnOffDelay => 5 sec,
              RestartDelay => 300 sec)
  port map(MaSw, HoTe, TeSe, MoSp, WaTe, FuFl, Co, MoSw,
           IgSw, OiVa, CiVa);

  MaSw <= heat, off after 1000 sec;
  HoTe <= HoTe + 1 after 10 sec when IgSw = onn and
                                     CiVa = openn else
          HoTe - 1 after 10 sec when HoTe > 60 else
          60 after 10 sec;
  WaTe <= WaTe + 1 after  10 sec when IgSw = onn else
          WaTe - 1 after 10 sec when WaTe > 60 else
          60 after 10sec;
  MoSp <= 0 after 1sec when MoSw = off else
          MoSp + 100 after 1 sec when MoSp < 1500  else
          1500 after  1 sec;
```

```
end a;

use work.all;
configuration c of tb is
  for a
    for all : whc_c
      use entity HeatingSystem(a);
      --| valentity;
    end for;
  end for;
end c;
```

9.4 Simulation Results

The errors in the VHDL description described here were actually made by the authors in writing the VHDL description and were caught during the VAL comparative simulation. Two assertion violations occur in the following simulation trace. The first, at 365 sec, occurs because the implementation state machine does not change to state **WAITING** as soon as the VAL specification requires. Examining the VHDL architecture for the water heater controller reveals that the process describing the state machine does not react when the implementation has been in the **DEACTIVATE** state for 5 seconds. The machine should change to **WAITING** 5 seconds after it changes to **DEACTIVATE**. Instead, no change in state occurs until a change on the inputs at time 370 wakes up the process. At that time, the process realizes that the 5 seconds has expired, and the state changes.

The second assertion violations occurs for a similar reason. In that case, the description is in state **WAITING** and fails to change to state **IDLE** on time.

Both assertion violations can be corrected by adding the signal indicating the delay to the sensitivity list of the VHDL process. The resulting process declaration becomes:

```
process(MasterSwitch, HomeTemp, TempSetting, MotorSpeed,
        WaterTemp, FuelFlow, Combustion,
        vhdl_state'stable(TurnOffDelay),
        vhdl_state'stable(RestartDelay))
```

TIME (FS)	MASW	MOSP	HOTE	WATE	MOSW	IGSW	OIVA	CIVA	VHDL_STATE	VALSTATEPORT
	SIGNAL NAMES									
0	HEAT	0	60	60	OFF	OFF	CLOSED	CLOSED	IDLE	IDLE
+1	HEAT				OFF	OFF	CLOSED	CLOSED	MOTORSTART	
+2					ONN					MOTORSTART
1		100								
2		200								
3		300								
4		400								
5		500								
6		600								
7		700								
8		800								
9		900								
10		1000	60	60						
+1									HEATING	
+2						ONN	OPENN			HEATING
11		1100								
12		1200								
13		1300								
14		1400								
15		1500								
16		1500								
20			60	61						
30				62						
40				63						
50				64						
60				65						
70				66						
80				67						
90				68						
100				69						
110				70						
120				71						
130				72						
140				73						
150				74						
160				75						
+1									CIRCULATING	
+2								OPENN		CIRCULATING

```
 170 |           61  76
 180 |           62  77
 190 |           63  78
 200 |           64  79
 210 |           65  80
 220 |           66  81
 230 |           67  82
 240 |           68  83
 250 |           69  84
 260 |           70  85
 270 |           71  86
 280 |           72  87
 290 |           73  88
 300 |           74  89
 310 |           75  90
 320 |           76  91
 330 |           77  92
 340 |           78  93
 350 |           79  94
 360 |           80  95
  +1 |                                        DEACTIVATE
  +2 |                     OFF CLOSED                     DEACTIVATE
 365 |
  +3 |                                                      WAITING
 365 | %VHDSIM-W-ASSERTV Assertion Violation after 3650000000000000001
     | f    at line 427 in design unit <<LMA.BOOK.WHC>>VALOUT_HEATINGS
     | YSTEM(STATEMONITOR)
     | incorrect output in state waiting
 370 |           79  94
  +1 |                                         WAITING
  +2 |                 OFF             CLOSED
 371 |       0
 372 |       0
 380 |           78  93
 390 |           77  92
 400 |           76  91
 410 |           75  90
 420 |           74  89
 430 |           73  88
 440 |           72  87
 450 |           71  86
 460 |           70  85
 470 |           69  84
 480 |           68  83
 490 |           67  82
 500 |           66  81
 510 |           65  80
 520 |           64  79
 530 |           63  78
 540 |           62  77
 550 |           61  76
 560 |           60  75
 570 |           60  74
 580 |               73
```

```
590 |              72
600 |              71
610 |              70
620 |              69
630 |              68
640 |              67
650 |              66
660 |              65
665 |
 +3 |                                                    IDLE
 +5 |                                                    IDLE
 +8 |                                                  MOTORSTART
665 | %VHDSIM-W-ASSERTV Assertion Violation after 665000000000000001
    | f    at line 299 in design unit <<LMA.BOOK.WHC>>VALOUT_HEATINGS
    | YSTEM(STATEMONITOR)
    | incorrect output in state motorstart
670 |              64
 +1 |                                      IDLE
680 |              63
 +1 |                                    MOTORSTART
 +2 |                 ONN
681 |     100
682 |     200
683 |     300
684 |     400
685 |     500
686 |     600
687 |     700
688 |     800
689 |     900
690 |     1000     62
 +1 |                                     HEATING
 +2 |                    ONN OPENN                         HEATING
691 |     1100
692 |     1200
693 |     1300
694 |     1400
695 |     1500
696 |     1500
700 |          60  63
```

Chapter 10

CPU Example

This example outlines a three level hierarchical design of a very simple CPU, and how annotations are used. The full details of the example are in Appendix B.

10.1 Requirements

10.1.1 Instruction level specification

The CPU operates on types bit and `bit_vector` (unconstrained array of bits).

The CPU has two data ports, one for input data (`d_i`), the other for output data (`d_o`), and one instruction port (`ir_i`). Those three ports are 16 bits wide. There is an address port (`addr_o`), 12 bits wide, and six control pins, each one bit wide. The output control pins are: `read_o`, `write_o`. The input control pins are the clock (`clk_i`), a reset pin (`rst_i`), a run pin (`run_i`), and the clear pin (`clr_i`).

The state of the CPU consists of a bank of four data registers, an instruction register, and an accumulator. The CPU can perform four instructions, load (`ld`), store (`st`), execute (`ex`) and idle (`id`). The format of each of the four instructions appears in figure 10.1.

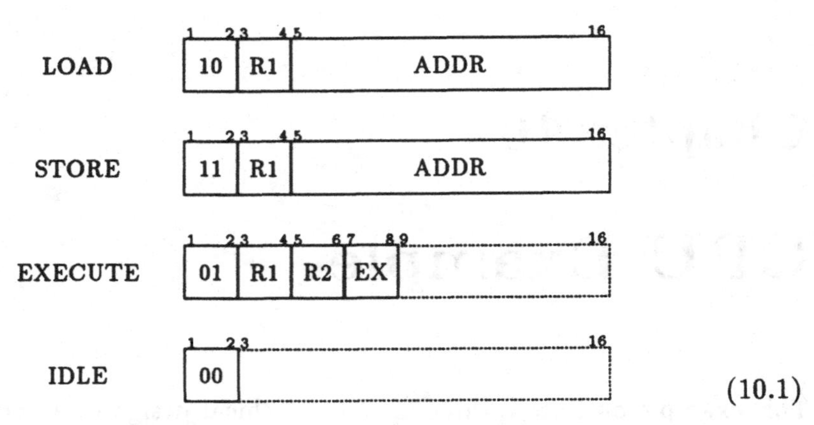

(10.1)

The first two bits of the instruction is the OP field. Load (OP = 10) takes the data from memory address ADDR and stores it in register R1. Store (OP = 11) takes the data from register R1 and stores it in memory at address ADDR. Execute (OP = 01) takes the data from register R1 and R2, performs a logical operation on those data, and returns the data in register R2 The operation performed depends on EX:

```
ex = 0     =>  r2 = r1 and r2
ex = 1     =>  r2 = r1  or r2
ex = 2     =>  r2 =    not r2
ex = 3     =>  r2 = r1 xor r2
```

All the control pins are active high (they take effect when their value is '1'. The reset pin asynchronously resets the CPU in its initial state (see definition of initial state below). While the run pin is low, the CPU once returned to its initial state, stays there. The clear pin resets the instruction register.

The CPU is a finite state machine. Figure 10.2 shows the state diagram of the CPU controller. The state diagram is not complete, in the sense that some transition conditions do not appear in the figure. The full description of state transitions appears right after the figure.

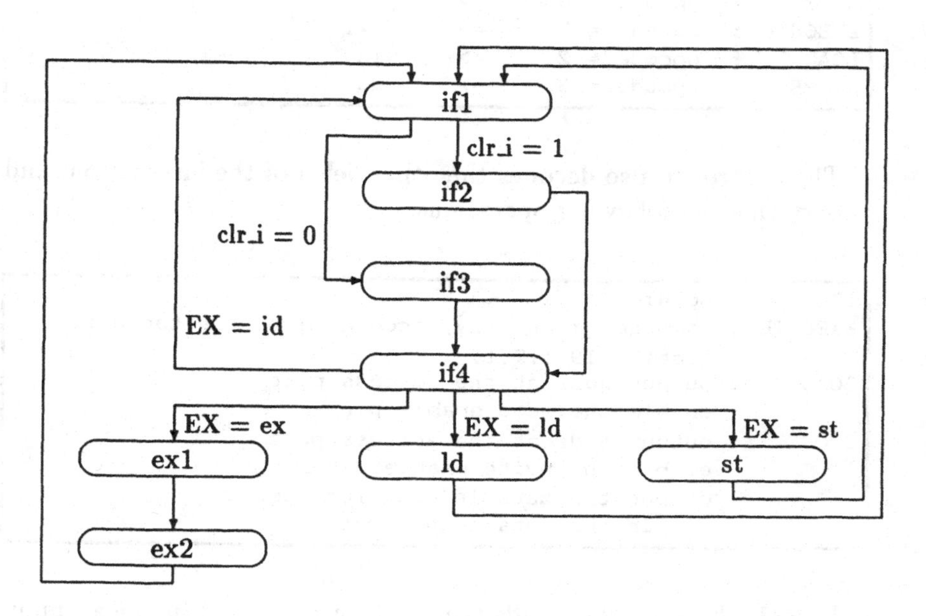

(10.2)

There are eight states in the CPU: instruction_fetch_1 (`if1`), instruction_fetch_2 (`if2`), instruction_fetch_3 (`if3`), instruction_fetch_4 (`if4`), execute_1 (`ex1`), execute_2 (`ex2`), load (`ld`) and store (`st`).

1. **if1** is the initial state of the state machine. When the `reset` pin is activated, the CPU state machine asynchronously returns to that state. If `clr_i` is active, the controller goes to state `if2`. If `clr_i` is not active and `run_i` is active, the controller goes to state `if3`.

2. In **if2**, the controller synchronously sets all bits of the instruction register to zero, and goes to `if4`.

3. In **if3**, the controller synchronously loads the instruction register from the instruction port, and goes to `if4`.

4. In **if4**, the controller decodes bit 1 and 2 (the opcode, OP) of the instruction. The following table shows to which state transition corresponds each opcode. The transitions occur only if the `run_i` pin is active. Otherwise, the controller stays in state `if4`.

```
IDLE     :  opcode = 0      =>      if1
EXECUTE  :  opcode = 1      =>      ex1
LOAD     :  opcode = 2      =>      ld
STORE    :  opcode = 3      =>      st
```

The controller also decodes the other fields of the instruction, and performs the following operations:

```
IDLE     : nothing
EXECUTE  : compute {r1 ex r2}, according to the table of
           section 19.1.2.1.
LOAD     : output addr at the address port
           enable the read_enable pin
STORE    : output addr at the address port
           enable the write_enable pin
           output the data in bank register r1
               in the data output port
```

5. In **ex1**, the controller loads the result of {r1 ex r2} in the accumulator, and goes to `ex2`.
6. In **ex2**, the controller transfers the value in the accumulator to bank register `r2`, and goes to state `if1`.
7. In **load**, the controller loads the data from the data input port in bank register r1, disables the read_enable pin and the address port, and goes to state `if1`.
8. In **store**, the controller disables the write_enable pin, the address port and the data port, and goes to state `if1`.

10.1.2 Register transfer level specifications

At the register transfer level, the CPU consists of a network of components, whose types are: arithmetic logic unit (`alu`), programmable logic array (`pla`), one output register (`ref`), two output registers (`reg`), and buffers (`buf`). A formal description of the network is available in section B.6. The following sections describe the specification for each of the component.

10.1.2.1 ALU

The ALU has two 16-bit inputs (`a_i` and `b_i`) and one 16-bit output (`c_o`, and one 2-bit operation selection output (`op_i`). The list of operation the ALU can perform is:

```
op = 00     =>     c = a and b
op = 01     =>     c = a  or b
op = 10     =>     c =   not b
op = 11     =>     c = a xor b
```

10.1.2.2 One output register

The one output register (REF) has a 16-bit input (`d_i`) and one 16-bit output (`d_o`). The control pins are a synchronous reset (`rst_i`), a clock (`clk_i`), a clock enable (`CE_i`), and an output enable (`OE_i`).

When the clock rises, if `CE_i` is active, the register loads one of two values: `'0'`, if `rst_i` is active, `d_i` otherwise. The output takes the value of the register state if `OE_i` is active, `'0'` otherwise.

10.1.2.3 Two output register

The two output register (REG) has a 16-bit input (`d_i`) and two 16-bit output (`d1_o` and `d2_o`). The control pins are a synchronous reset (`rst_i`), a clock (`clk_i`), a clock enable (`CE_i`), and two output enables (`OE1_i` and `OE2_i`).

When the clock rises, if `CE_i` is active, the register loads one of two values: 0, if `rst_i` is active, `d_i` otherwise. Output `d1_o` takes the value of the register state if `OE1_i` is active, `'0'` otherwise. Output `d2_o` takes the value of the register state if `OE1_i` is active, `'0'` otherwise.

10.1.2.4 Buffer

The buffer (BUF) has a 16-bit input (`d_i`) and a 16-bit output (`d_o`). There is one control pin, an output enable (`OE_i`).

The output takes the value of the input if `OE_i` is active, `'0'` otherwise.

10.1.2.5 Programmable Logic Array

The Programmable Logic Array (PLA) is the controller for the CPU. It controls the sequencing operations described in section 10.1.1. It makes to activate the right enable pins at the right moments.

The input pins of the PLA are: the OP, R1 and R2 fields of the instruction (**op_i**, **r1_i**, and **r2_i**), the **clr_i**, **run_i**, **rst_i** pins coming from the CPU interface, and the clock.

The outputs are: the instruction register clock enable (**irCE_o**), the accumulator register clock enable (**accCE_o**), the four bank registers clock enables (**reg0CE_o** to **reg3CE**), the four bank registers first output enables (**reg0OE1_o** to **reg3OE1_o**), the four bank registers second output enables (**reg0OE2_o** to **reg3OE2_o**), the accumulator register output enable (**accOE_o**), the input data buffer output enable (**dinOE_o**), the instruction register reset (**irRST_o**), the address buffer output enable (**addrOE_o**), the read enable buffer output enable (**readOE_o**), and the write enable buffer output enable (**writeOE_o**).

The PLA has eight states, named **if1**, **if2**, **if3**, **if4**, **ex1**, **ex2**, **ld**, and **st**, corresponding to the eight states of figure 10.2. The state transitions are also explained in section 10.1.1.

To each state corresponds a set of active output pins:

```
if1 : accCE
if2 : irRST, irCE
if3 : irCE
if4 :
ex1 : reg0OE1_o if r1_i = 00,
      reg1OE1_o if r1_i = 01,
      reg2OE1_o if r1_i = 10,
      reg3OE1_o if r1_i = 11,
      reg0OE2_o if r2_i = 00,
      reg1OE2_o if r2_i = 01,
      reg2OE2_o if r2_i = 10,
      reg3OE2_o if r2_i = 11
ex2 : reg0CE_o if r2_i = 00,
      reg1CE_o if r2_i = 01,
      reg2CE_o if r2_i = 10,
      reg3CE_o if r2_i = 11,
ld  : reg0CE_o if r1_i = 00,
      reg1CE_o if r1_i = 01,
```

```
          reg2CE_o if rl_i = 10,
          reg3CE_o if rl_i = 11,
          readOE_o,
          addrOE_o
st      : reg0OE1_o if rl_i = 00,
          reg1OE1_o if rl_i = 01,
          reg2OE1_o if rl_i = 10,
          reg3OE1_o if rl_i = 11,
          writeOE_o,
          addrOE_o
```

10.1.2.6 Architecture

Figure 10.1 is a graphic representation of the CPU architecture. The flow of data across the CPU components is as follows: the bank of four registers (`R(0)`, `R(1)`, `R(2)`, and `R(3)`) stores operands to be processed by the logic unit `A`; an accumulator `ACC` stores temporarily the result output by the logic unit, before it is transferred to the register bank; two buffers, `DIB` (data in buffer) and `DOB` (data out buffer), transfer data from the outside world to the register bank, and vice versa; an instruction register `IR`, stores an instruction (`ins` parameter of the `Instr` action) for the duration of that instruction; an address buffer `AEB`, issues addresses to the memory where from the CPU fetches data; and finally the PLA `C` controls the behavior of the registers and buffers. (for example, the controller `Accoe_o` pin is connected to the `Ce_i` pin of the `Acc` register).

10.1.3 Gate level specifications

Each of the register transfer level components is described in terms of gates (VHDL predefined operators **and**, **or**, **not**). The VHDL descriptions for the ALU, REF, REG and BUF components first describe the 16-bit components in terms of one bit components (see sections B.2, B.13, B.15, B.5, and B.4), then describe the one bit components in terms of gates (see sections B.1, B.12, B.14, B.3). The VHDL description for the PLA consists of three one bit registers and gates (see section B.11).

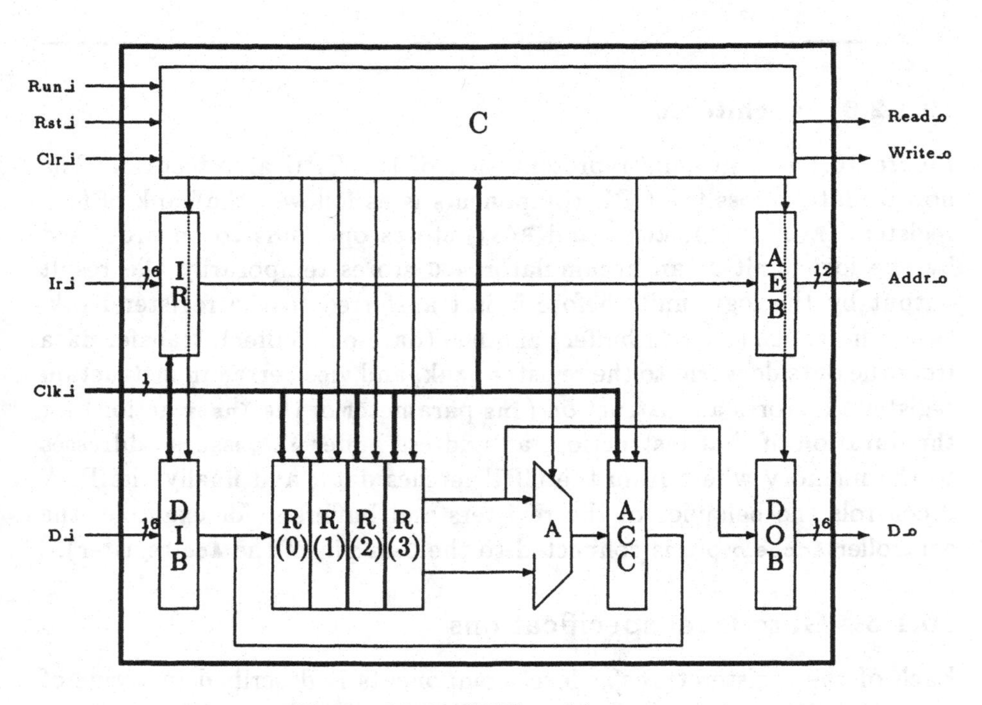

Figure 10.1: CPU architecture

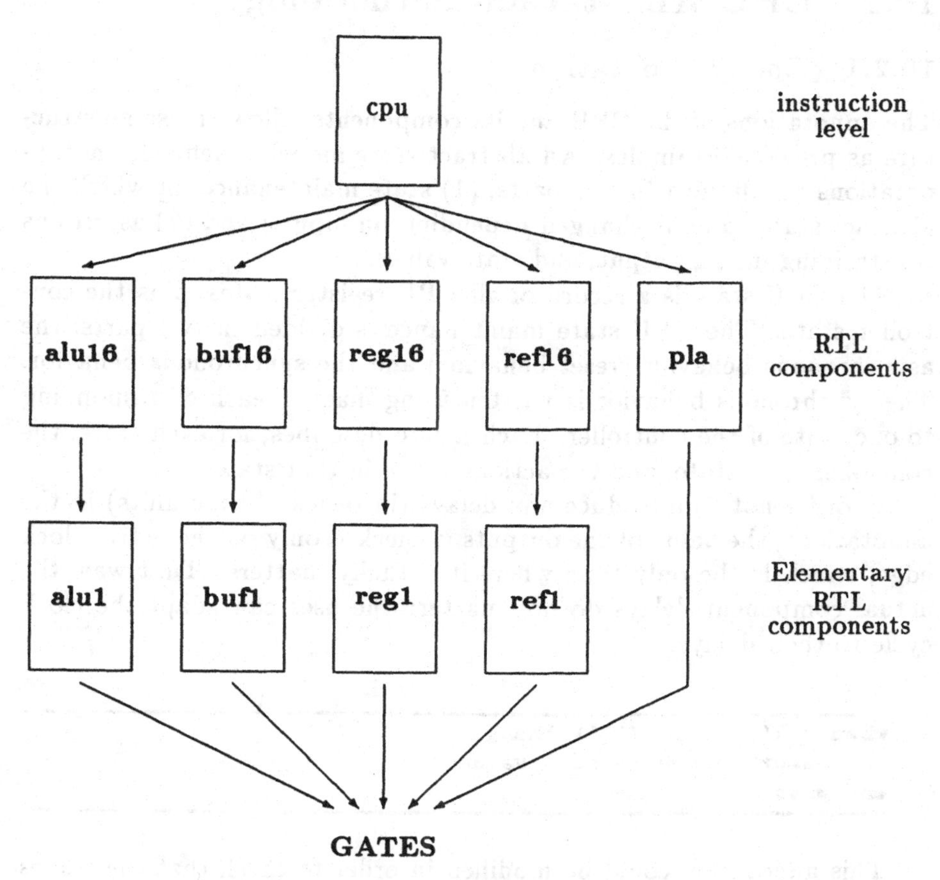

Figure 10.2: CPU hierarchy of components

10.1.4 Hierarchy of components

Figure 10.1.4 gives a graphic representation of the dependencies between components.

10.2 CPU Annotation methodology

10.2.1 Entity annotation

The annotations of the CPU and its components follow the same structure as previous examples. An abstract state model is defined, and annotations are divided in two parts, (1) state maintenance, in which the abstract state value is changed depending on inputs; and (2) assertions constraining input, output, and state values.

The CPU state is a record of all CPU register states, plus the controller state. The CPU state maintenance is divided in two parts, the asynchronous behavior (reset behavior) and the synchronous behavior. The synchronous behavior is written using macros, each corresponding to one state of the controller. Each macro describes, for each state, the controller next state, and the actions taken in that state.

In order not to introduce any delays (in terms of time units) in the annotations, the value of the outputs is checked only on the active clock edge, which is the only time where it actually matters. That way, the actual component delays do not matter: the user can adapt the clock cycle to their delays.

```
when clk'changed ('1') then
    assert output = some_value ;
end when ;
```

This annotation could be modified in order to check that the values are stable during the interval starting `setup` time units before the clock edge, and `hold` time unit after the clock edge:

```
when clk'changed ('1') then
    assert output = some_value
        during [-setup, hold] ;
end when ;
```

The output assertions are divided two parts, one corresponding to each state (`load` and `store`) during which the outputs are active.

```
when clk'changed ('1') then
    when state = ld then
        first_output = active ;
        second_output = some_value ;
    end when ;
    when state = st then
        second_output = some_other_value ;
        third_output = active ;
    end when ;
 end when ;
```

There is an alternative way of structuring the output assertions, with one guarded statement per output. This way, one can constrain not only when the outputs must be active, but also the values the output takes when it is not active.

```
when clk'changed ('1') then
    when C1 and C2 and C3 then
        first_output = some_value ;
    elsewhen C4 and C5 then
        first_output = some_other_value ;
    else
        first_output = default_value ;
    end when ;
    when C6 then
        second_output = some_value ;
    else
        second_output = default_value ;
    end when ;
    ...
end when ;
```

10.2.2 Mapping

The architecture annotations allow the designer to check the architecture. Architecture annotations have visibility to both the entity state, and the architecture component states. That way, the designer can check the consistency between the states at different levels of his design hierarchy.

The CPU architecture annotations define the mapping between the component states and the entity state. Also, entity annotations are used to constrain states and output values, but only when the active clock edge occurs. A typical schematic example of entity annotations is:

```
when clk'changed ('1') then
    assert state.st = pla.state ;
end when ;
```

10.3 VHDL description

The VHDL description for the CPU is available in appendix B. The appendix lists the components alphabetically.

Part III

The VAL Language Reference Manual

Chapter 11

Lexical Elements

Lexical elements of VAL closely follow those of VHDL. The text of annotations is a sequence of lexical elements, each of which is composed of characters. This chapter gives the rules of composition. Since lexical elements of VAL closely follow those of VHDL, we describe only the differences. (See Appendix A of the VHDL LRM [25] for details of VHDL lexical elements.)

11.1 Character Set

VAL uses the same character set as VHDL. (See Section 13.1 of the VHDL LRM [25] for the VHDL character set.)

```
graphic_character ::= ...(same as VHDL)
```

11.2 Lexical Elements, Separators, and Delimiters

VAL has the lexical elements, separators, and delimiters of VHDL (see Section 13.2 of the VHDL LRM [25]) with the following additional delimiters:

```
delimiter : ->, <-
name      : drive

delimiter : --|
name      : annotation
```

11.3 Identifiers

VAL recognizes the same class of identifiers as VHDL. (See Section 13.3 of the VHDL LRM [25].)

```
identifier ::= ....(same as VHDL)
```

11.4 Literals

VAL recognizes the same literals as VHDL – abstract literals representing real and integer numbers, character literals, string literals, and bit string literals. (See Sections 13.4, 13.5, 13.6, and 13.7 of the VHDL LRM [25].)

```
abstract_literal ::= (same as VHDL)
character_literal ::= (same as VHDL)
string_literal ::= (same as VHDL)
bit_string_literal ::= (same as VHDL)
```

11.5 Comments

A comment starts with two adjacent hyphens; these may be followed immediately by any character except a vertical bar ('|'). Comments extend to the end of the line. A comment can appear on any line of a VAL description. Comments have no influence on the semantic interpretation of the description. Their sole purpose is the enlightenment of the human reader.

Example:

```
-- This text is a comment
-- ----- The first two hyphens start the comment
--| This is not a comment.  It is a VAL annotation
```

11.6 Annotations

Each line of annotations begins with the annotation indicator **--|** (two hyphens and a vertical bar). Thus, annotations are distinguished from regular comment text, and also distinguished from pure VHDL source code.

Example:

```
--| assert (X and Y) = Z ;
```

According to the VHDL lexical rules, annotations are interpreted as comments. In BNF notation it is difficult to express that every line of an annotation must begin with the **--|** annotation indicator. By convention, if the right hand side of a BNF clause starts with **--|**, it means that all lines matching that rule in a VHDL file must start with the **--|** annotation indicator.

11.7 Reserved Words

VAL reserved words consist of all the VHDL reserved words (see Section 13.9 of the VHDL LRM [25]) and the additional keywords listed below. These reserved words have special meaning in VAL. For readability, reserved words appear in lower case boldface in examples in this text. Reserved words must not be used as declared identifiers in annotations.

assume	**macro**	**valarchitecture**
during	**model**	**valconfiguration**
elsewhen	**sometime**	**valentity**
eventually	**state**	**within**
finally		

11.8 Allowable Replacements of Characters

VAL allows replacement of the vertical bar, sharp, and quotation characters as per VHDL. (See Section 13.10 of the VHDL LRM [25] for definitions and recommendations for the use of these replacements.)

11.9 BNF Notation

The conventions described in the preface of the VHDL LRM [25] all apply. The two differences come from the fact that the symbols |, [, and] are used both in the description of the context free syntax of the language and in the language itself. The following convention distinguishes the two uses: the quoted symbols '|', '[', and ']' refer to the symbols as they appear in the language, and the plain symbols |, [, and] are part of the BNF notation.

Chapter 12

Design Units

VAL annotations are classified into three categories according to the design unit in which they appear. Entity annotations appear in an entity declaration, architecture annotations appear in an architecture body, and configuration annotations appear in a configuration declaration. This chapter presents an overview of these three different categories of annotations.

12.1 Entity Annotations

Entity annotations appear in the VHDL entity declaration. The VHDL entity declaration consists of the entity header, the entity declarative part, and the entity statement part. Entity annotations may appear in the entity declarative part and the entity statement part according to the following syntax:

```
Annotated_VHDL_entity_declaration ::=
    entity identifier is
        VAL_VHDL_entity_header
        VAL_VHDL_entity_declarative_part
  [ begin
      VAL_VHDL_entity_statement_part ]
    end [ entity_simple_name ] ;

VAL_VHDL_entity_header ::=
```

```
    entity_header
  [ VAL_state_model_declaration ]

entity_header ::= ...(same as VHDL)

VAL_VHDL_entity_declarative_part ::=
  { VAL_VHDL_entity_declarative_item }

VAL_VHDL_entity_declarative_item ::=
    --| VAL_entity_declarative_item
  | entity_declarative_item

entity_declarative_item ::= ...(same as VHDL)

VAL_VHDL_entity_statement_part ::=
  { VAL_VHDL_entity_statement }

VAL_VHDL_entity_statement ::=
    --| VAL_entity_statement
  | entity_statement

entity_statement ::= ...(same as VHDL)
```

There are three general classes of entity annotations:

1. **VAL state model:** The state model declaration appears immediately after the VHDL entity header. It is optional. The state model is described in detail in Chapter 13.

2. **VAL declarations:** VAL declarations appear in the entity declarative part, mixed with VHDL declarations. VAL declarations are described in detail in Chapter 14.

3. **VAL statements:** VAL statements appear in the entity statement part of the entity declaration. The syntax and semantics of VAL statements are described in detail in Chapter 16.

VAL annotations in the entity declaration obey the linear visibility rules of VHDL. An annotation has visibility over all the preceding VAL and VHDL statements. However, VAL annotations are invisible to the VHDL statements. A VHDL statement cannot see any preceding VAL statements. For more details on the visibility rules, refer to section 19.2.

Example:

```
entity foo is
    port (...);
    --| state model is ...;
    type three_color is (red, blue, yellow);
    --| subtype three_color_2 is three_color;
    subtype traffic_light is three_color_2;
    ...
end foo;
```

In this example, **three_color** is visible to the subtype annotation of **three_color_2**. However, the VAL subtype annotation **three_color_2** is not visible to the VHDL type declaration **traffic_light**. Thus, the VHDL type declaration **traffic_light** will result in an error during semantic analysis.

Although the syntax allows VAL annotations to be dispersed in the VHDL source code, we suggest the following annotation style for readability:

1. VAL declarations, except the state model declaration, should be located after all the VHDL declarations.

2. VAL syntax allows both VHDL entity statements and VAL statements to coexist in the entity statement part starting right after the keyword **begin** in the VHDL entity declaration. In such a case, it is recommended that all the VAL statements are grouped and located before the VHDL entity statements.

12.2 Architecture Annotations

Architecture annotations can appear only in the concurrent part of a VHDL architecture. Thus, architecture annotations can appear only inside VHDL blocks, and are not allowed inside VHDL processes. There are three places in architectures where annotations may appear:

- At the top level:

```
Annotated_architecture_body ::=
    architecture identifier of entity_name is
        VAL_VHDL_architecture_declarative_part
    begin
        VAL_VHDL_architecture_statement_part
    end [ architecture_simple_name ] ;

VAL_VHDL_architecture_declarative_part ::=
   { VAL_VHDL_architecture_declarative_item }

VAL_VHDL_architecture_declarative_item ::=
        --| VAL_architecture_declarative_item
      | architecture_declarative_item

architecture_declarative_item ::= ...(same as VHDL)

VAL_VHDL_architecture_statement_part ::=
  { VAL_VHDL_architecture_statement }

VAL_VHDL_architecture_statement ::=
        --| VAL_architecture_statement
      | architecture_statement

architecture_statement ::= ...(same as VHDL)
```

- **In a block statement:**

```
Annotated_block_statement ::=
    block_label :
        block [ (guard_expression) ]
            block_header
            VAL_VHDL_block_declarative_part
        begin
            VAL_VHDL_block_statement_part
        end block [ block_label ] ;

block_header ::= ...(same as VHDL)

VAL_VHDL_block_declarative_part ::=
  { VAL_VHDL_block_declarative_item }

VAL_VHDL_block_declarative_item ::=
```

```
      --| VAL_architecture_declarative_item
    | block_declarative_item

block_declarative_item ::= ...(same as VHDL)

VAL_VHDL_block_statement_part ::=
    { VAL_VHDL_block_statement }

VAL_VHDL_block_statement ::=
      --| VAL_architecture_statement
    | block_statement

block_statement ::= ...(same as VHDL)
```

• **In a component declaration:**

```
component_declaration ::=
    component identifier
      [ generic ( local_generic_list ) ]
      [ port ( local_port_list ) ]
      [ --| VAL_state_model_declaration ]
    end component ;
```

There are three kinds of VAL architecture annotations:

1. **VAL declarations:** VAL declarations appear in the architecture declarative part, mixed with VHDL declarations. VAL declarations are described in detail in Chapter 14.

2. **VAL statements:** VAL statements appear in the architecture statement part of the architecture declaration, mixed with VHDL statements. VAL statements are described in detail in Chapter 16.

3. **Component state model declarations:** The state model declaration inside a component is optional. It appears in a VHDL component declaration, right after the port declarations. The VAL state model declaration is described in detail in Chapter 13.

The entity state declared in entity annotations is visible to architecture annotations. Visibility rules of the entity state to the architecture annotations are the same as those of entity ports.

12.3 Configuration Annotations

VHDL configurations specify binding of instances of components with a specific entity declaration and architecture. These configurations define a design hierarchy that will be simulated. VHDL configurations are expressed in (1) a separate VHDL configuration declaration, or (2) a configuration specification declared in the VHDL architecture.

The information about inclusion of VAL annotations in the design hierarchy for simulation is expressed in VAL *configuration annotations*. VAL configuration annotations allow the user to specify which annotated VHDL entities and architectures are included for self-checking simulation.

Configuration annotations are also used to resolve the difference between the local state model of a component declaration and the state model of the entity declaration that is instantiated as the component. Those annotations are called *state model map*.

Details of configuration annotations and state model map are described in Chapter 18.

Chapter 13

State Model

The entity state is an object declared through a VAL entity state model declaration. The entity state has memory capability. Once a value is written to the entity state, it is maintained until it is overridden by a new assignment. The state value can be read by any VAL process.

The entity state is visible only to VAL annotations; it is not visible to VHDL statements. Entity annotations may both read and modify the state. The drive process (see Section 16.2) is the only process which can affect a change in the entity state. All other processes may read the state exactly as if it were a VHDL signal. Architecture annotations can access the entity state on a read-only basis.

13.1 State Model Declaration

The syntax of the state model declaration is:

```
state_model_declaration ::=
   state model [ identifier ] is
        subtype_indication := expression ;
subtype_indication ::= ...(same as VHDL)
```

A stàte model can appear in an entity declaration immediately after the port declaration (*entity state*), or inside an architecture in a component declaration immediately after the port declaration (*component state*). The subtype indication and the initial value are required. The state model type can be a simple predefined VHDL data type such as **BIT**, **BOOLEAN**, and **INTEGER**, or a user defined VHDL type such as an enumeration, record, or array type.

The **state model** may be optionally given a name. The identifier following **state model** declares the name of the entity state. If the identifier is omitted, **STATE** is used as the default entity state name. The entity state is referred to by **entity_name.state_name** or simply **state_name**, where **state_name** is an identifier. When the entity state of a component is referred to in an architecture, the entity state is denoted by **component_label.state_name**. The optional entity state name is usually used to avoid a name conflict when the name **state** has already been defined by the user in VHDL.

The default value defines the initial value of the entity state. It is required.

Example:

> If **CONTENT** is declared as the entity state name of entity **BINGO**, the entity state can be referred to by either **BINGO.CONTENT** or **CONTENT** in annotating the entity and its architectures.

Example:

> If an entity state name is omitted in the above example, the entity state is referred to by either **BINGO.STATE** or **STATE** in annotating the entity and its architectures.

Example:

> When an entity is instantiated as a component with the component label **COUNTER** in an architecture, the entity state of the component is referred to by **COUNTER.STATE** if the default state name is used.

13.2 State Model Type

There are three different approaches to declaring state models in VAL. They will be described in this section.

Predefined VHDL Types as State Model

The first approach is to declare a state model as a predefined VHDL type such as **INTEGER**, **REAL**, **NATURAL**, **BIT**, **BOOLEAN**, and **CHARACTER**.

Example:

```
--| state model is BIT := '0' ;
```

The entity state is of VHDL type **BIT**.

Operations on an entity state are the operations provided by the VHDL type used as the state model.

User-Defined State Model Types

The second approach is to declare a user-defined type, and use it as a state model. User-defined types can be VHDL enumeration, array, or record types.

A user-defined type used as a state model is declared in a separate package and imported into the VHDL entity declaration. In that respect the state model is similar to ports, for which a user-defined type must be declared in a separate package to allow the declaration to be visible across design units.

Example:

When the entity state of entity **Shift_register** has type **Register_array_type** that is assumed to be declared in separate package **Type_declaration**, the state model can be declared as follows:

```
use Type_declaration.all;  -- context clause
entity Shift_register is
    port ( ... );
    --| state model is Register_array_type := initial_value;
end Shift_register;
```

Example:

If the package **Type_declaration** is imported only for the type of an entity state model, the use clause can appear as an annotation.

```
entity Shift_register is
    port ( ... );
    --| use Type_declaration.Register_array_type;
    --  Context clause is only for
    --  name Register_array_type
    --| state model is Register_array_type := initial_value;
end Shift_register;
```

Example:

If the entity **traffic_light_controller** is specified as a finite state machine with four distinct states, **Highway_Green**, **Highway_Yellow**, **Farmroad_Green**, and **Farmroad_Yellow**, the state type of **traffic_light_controller** is declared as the following enumeration type, which is used as a state model:

```
package type_decl is
   --| type tlc_state_type is
   --|      (Highway_Green, Highway_Yellow,
   --|       Farmroad_Green, Farmroad_Yellow);
end type_decl;

use type_decl.all;
entity traffic_light_controller is
    ...
    --| state model is tlc_state_type := Highway_Green;
    ...
end example;
```

Example:

If the entity state of a 16-bit shift register is modeled as an array of bits, the type is declared first in a package and then imported into the entity declaration.

```
package type_decl is
    --| type word is BIT_VECTOR (0 to 15);
    ...
end type_decl;
```

```
use type_decl.all;
entity shift_register is
    ...
    --| state model is word := BIT_VECTOR'(X"0000");
    ...
end example;
```

Example:
If the entity state of a hardware stack is modeled as a **TOS_pointer**, indicating the top of stack, and an array of 16 bit registers whose array size is given by **Stack_size**, it is declared as follows:

```
package type_decl is
    --| constant stacksize : integer := 32 ;
    --| type word is bit_vector (0 to 15) ;
    --| type stack_range is range 1 to Stack_size;
    --| type word_array is array (stack_range)
    --|                 of word;
    --| type stack_type is record
    --|      Register_array: word_array;
    --|      TOS_pointer:   stack_range;
    --| end record;
    ...
end type_decl;

use type_decl.all;
entity stack is
    ...
    --| state model is stack_type := initial_value;
    ...
end example;
```

Abstract Data Type as State Model

The third approach is to use *abstract data types* [10] as an entity state model. The concept of an abstract data type has been developed in software engineering, and is supported as a major abstraction mechanism in most modern programming languages such as Ada and C++. An abstract data type consists of a data type and a set of functions applying to the data type. Access to and manipulation of the object with this data type is possible only by calling these functions.

An abstract data type is defined by declaring a data type and a set of functions in a package. The abstract data type is imported into the entity annotation by a context clause. When an entity state is modeled as an abstract data type in a VHDL entity declaration, the entity state can be modified only by calling a function provided by the abstract data type.

Abstract data types may provide various data abstractions such as queues, stacks, lists, and sets. When abstract data types are available in the VHDL design library, hardware designers can build VAL behavior specifications more easily and quickly on top of library abstractions by utilizing these data types.

Example:

Assume that the state of hardware entity **Context_Switcher** is modeled as an abstract data type **Stack**. The abstract data type **Stack** is defined in a package as follows:

```
-- package specification of the abstract data type Stack
package stack_package is
    type item is BIT_VECTOR (0 to 15);
    type register_array is array (0 to 31) of item;
    type stack is record
        top, bottom: integer;
        data: register_array;
    end record;
    function push(S: stack; data: item) return stack;
    function pop(S: stack) return stack;
    function top(S: stack) return item;
    -- The above functions are implemented in the package body
    -- which is usually hidden from the user of the package.
end stack_package;
```

```
-- VHDL entity declaration with annotations
--| use stack_package.all;
    -- context clause to import the stack package
entity Context_Switcher is
    port ( ... );
```

```
    -- model stack as an entity state.
    --| STATE model is stack := initial_value;
begin
    -- STATE is manipulated by
    -- functions push, pop and top
    ...
end Context_Switcher;
```

Commentary on local variables:

VAL does not support declaration of local variables and signals. The reason is that in conventional hardware description languages the use of local variables (or signals) has two major purposes: one is to represent the internal state of the hardware, and the other is for programming convenience such as loop parameters and temporary variables. In VAL, the former use of local variables is superseded by the entity state concept, and the latter use is restricted to loop parameters in the generate statement. Thus, declarations of local variables are no longer required in VAL.

Chapter 14

Declarations

VAL declarations may appear as annotations in the declarative part of VHDL entities and architectures. Types, subtypes, constants, aliases, use clauses, assumptions, and macros can be declared as VAL annotations in entities.

```
VAL_entity_declarative_item ::=
        type_declaration
      | subtype_declaration
      | constant_declaration
      | alias_declaration
      | use_clause
      | assume_declaration
      | macro_declaration
```

Aliases, use clauses, macros, VAL state model maps and VAL binding annotations can be declared as annotations in architectures.

```
VAL_architecture_declarative_item ::=
        alias_declaration
      | macro_declaration
      | use_clause
      | conf_state_model_map_annotation
      | conf_VAL_binding_specification
```

See Section 19.2 for a description of the visibility rules applying to VAL declarations.

14.1 Types, Subtypes, Constants, Aliases and Use Clauses

A type is defined by a set of values and a set of operations. VHDL provides four classes of types: scalar, composite, access, and file. VAL annotations can use only scalar and composite types and their associated operations. Subtypes may also be declared in VAL.

VAL statements are allowed only in the concurrent part of the VHDL architecture and entity declaration. Access and file types are supported only in the sequential part of VHDL. Thus, access and file types cannot be used in VAL annotations.

The following types may be used in VAL annotations:

- **CHARACTER**, **BIT**, **BOOLEAN**, and **SEVERITY_LEVEL** as predefined enumeration types. In contrast to VHDL, the default severity level of VAL is **WARNING**.
- User-defined enumeration types.
- Integer types.
- Physical types.
- **TIME** as a predefined physical type. While VHDL requires that values for type **TIME** should be non-negative, VAL allows both negative and positive values including zero.
- Floating point types.
- Array types.
- **STRING**, **BIT_VECTOR** as predefined array types.
- Record types.

Refer to Chapter 3 of the VHDL LRM [25] for the definition of each VHDL type and its operations. VAL uses the same rules as VHDL for declaration of type, subtype, constant, alias, and use clause.

- Type declaration - See Section 4.1 of the VHDL LRM [25].
- Subtype declaration - See Section 4.2 of the VHDL LRM [25].
- Constant declaration - See Section 4.3.1.1 of the VHDL LRM [25].
- Alias declaration - See Section 4.3.4 of the VHDL LRM [25].

- Use clause - See Section 10.4 of the VHDL LRM [25].

VAL declarations of assumptions, state models, and macros will be described in this chapter. VAL declarations appearing in the VHDL configuration specification will be described in Chapter 18.

14.2 Assumptions

The assume declaration is used to specify constraints on generic parameters, constants, and in general any static expression. See Section 7.4 of the VHDL LRM for a complete definition of a static expression.

```
assume_declaration ::=
    assume boolean_expression
  [ else ]
      [ report expression ]
      [ severity expression ] ;
```

The expression following the keyword **assume** should be a static expression of type **BOOLEAN**. When constraints cannot be static boolean expressions, they can be specified using VAL assertion statements instead of VAL assume declarations.

Assume declarations constrain the use of a VHDL entity and its architectures. Constraints stated in assume declarations should be respected by the user of a VHDL entity and its architectures. In other words, assumptions declared in the entity have to be satisfied whenever the entity is instantiated and used as component of a higher level architecture. If an assume declaration specifies a condition on generic parameters, it has to be satisfied by actual values for those generic parameters whenever the corresponding entity is instantiated.

To the implementor of an entity, assume declarations are conditions that can be assumed in building the architecture of the entity. The implementor can assume these conditions to be always observed by the user.

When an assume declaration is violated during elaboration of an entity, it is reported as an assertion violation with the specified message. The severity level specifies the action that has to be taken when the assumption is violated. Syntax and semantics of the report and severity clauses in the assume declaration follow the rules applying to report and severity clauses of **assert** statements (described in Section 16.1).

Example:

```
entity D_flipflop is
    generic (setup, hold, delay : TIME);
    port (clk, D : in BIT;
          Q, Qbar: out BIT);
    --| assume (delay >= hold) else
    --|     report "Error in generic constant";
end D_flipflop;
```

The assume declaration annotated in the above **D_flipflop** example constrains every instantiation of entity **D_flipflop** so that the delay time is greater than or equal to the hold time.

14.3 Objects

We distinguish VHDL objects from VAL objects. All the objects declared in the VHDL declarations such as ports and local signals are called VHDL objects (see Section 4.3 of the VHDL LRM [25] for definition of VHDL objects). Objects introduced by annotation declarations are called VAL objects. VAL objects comprise entity states and loop parameters. The entity state is the object declared through VAL state model declaration. The loop parameter is a local variable declared implicitly in the VAL generate statement. The scope of loop parameters are limited to the statements appearing within the generate statement (See section 16.5).

VHDL objects are not allowed to be modified by annotations. Annotations can access VHDL objects on a read-only basis.

14.4 Macros

A macro defines a parameterized template for a block of text and associates a name with the template. Wherever a macro is called in the annotation, the macro is expanded textually. The macro facility is useful when the same annotation text is replicated in multiple places or when a long portion of annotations needs to be abstracted and represented by a name. Unlike computational abstraction provided by procedures in programming languages, VAL macros provide a textual abstraction facility. By using macros annotations can be shortened considerably, which improves the readability and writability of annotations.

Macros are declared in the declarative part of annotations or in a separate package according to the following syntax.

```
macro_declaration ::=
    macro_specification is
        macro_body
    end simple_name;

macro_specification ::=
    macro simple_name [ ( macro_parameter_list ) ]
simple_name ::= ...(same as VHDL)
macro_parameter_list ::= identifier_list
identifier_list := ...(same as VHDL)

macro_body ::= VAL_statement_list
VAL_statement_list ::= { VAL_statement }
```

The specification of a macro declares its name and the optional parameter list. The macro name is an identifier.

The macro parameter list defines a list of formal parameters of the macro. These formal parameters are identifiers. When formal parameters are declared in the macro specification, their scopes are limited to the corresponding macro body. The body of a macro is a list of statements.

When a macro is called, it will be replaced by the corresponding macro body at compilation time. Parameters in a macro body text will be substituted by corresponding arguments when the macro call is expanded. The arguments in the macro call must be valid expressions. (See Section 16.6 for macro call.) Macros cannot be defined recursively. In other words, a macro cannot call itself, either directly or indirectly.

Example:

```
--| macro check_state is
--|     assert STATE.status = normal
--|         report "Emergency state"
--|         severity failure;
--| end check_state;
```

The above example can be generalized by introducing parameters **condition** and **message**:

```
--| macro check_state (condition, message) is
--|    assert STATE.status = condition
--|        report message
--|        severity failure;
--| end check_state;
```

Example:

The following macro definition is incorrect because it is defined recursively. Macro **check_status** calls itself inside its body.

```
--| macro check_status is
--|     check_interrupt;
--|     check_status;
--| end check_status;
```

Example:

The following is an example of mutually recursive macro definitions. Macro **check_status** and macro **check_interrupt** call each other, and are thus illegal:

```
--| macro check_status is
--|     ...
--|     check_interrupt;
--| end check_status;

--| macro check_interrupt is
--|     ...
```

```
--|      check_status;
--| end check_interrupt;
```

Chapter 15

Names and Expressions

Names and expressions of VAL closely follow those of VHDL. (Refer to Chapters 6 and 7 of the VHDL LRM [25].) VAL augments the expressions of VHDL with new features to describe timing behavior:

- *Timed expressions* — Expressions may be associated with time. Thus, expressions are evaluated with time-dependent values of subexpressions and objects such as ports and signals.
- *Time qualified boolean expression* — A boolean expression is qualified so that it is evaluated over a given time interval.

VAL expressions are extended from VHDL expressions as follows:

```
extended_expression ::=
    ...VHDL_expression ...
  | timed_expression
  | time_qualified_boolean_expression
```

Extended expressions can be nested. So, the following modification is necessary to the VHDL syntactic category primary:

```
primary ::=
      ...VHDL_primary ...
    | ( extended_expression )
```

15.1 Timed Expressions

```
timed_expression ::=
      ... VHDL_expression ...
    | expression '[' time_expression ']'
```

In the above syntax, '[' and ']' are the left and right square bracket characters. These quoted square brackets are used in order to distinguish them from square brackets denoting an optional part in the BNF notation. Refer to Section 2.1.4.1 for an informal description of the semantics of timed expressions.

The time expression must return a value with VHDL physical type **TIME** as the evaluation result. VAL time expressions should have time units such as ps (picosecond), ns (nanosecond), ms (millisecond), sec (second), and min (minute). However, VAL time expressions can have negative values unlike VHDL time expressions that have only positive time values. VAL is based on the concept of relative time that allows the reference time to be relative rather than absolute. Negative time means a reference to the value of a signal or expression in the past. Positive time refers to the value of a signal or expression in the future.

Example:

```
X[0 ns]
X[5 ms]
X[-4 sec]
X[t + 3 ns]
```

In the above example, **X[0 ns]** means the value of **X** at the current point in time. Thus, **X[0 ns]** has the same meaning as **X**. **X[5 ms]** means the value of **X** at 5 ms in the future while **X[-4 sec]** means the value of

`X` at 4 sec in the past. The value of `X[t + 3 ns]` is the value of `X` at the point in time given by the value of `(t + 3 ns)` evaluated at the current point in time.

A timed expression is also used to specify the time associated with an expression. The time associated with a timed expression means the time at which the expression is evaluated.

Example:

```
(X + Y)[-5 ns]
      -- expression (X + Y) 5 ns in the past.
(X + Y)
      -- expression (X + Y) at the current time.
(X + 2)[3 ns]
      -- expression (X + 2) 3 ns in the future.
```

The timing operator can be distributed over simple mathematical expressions while preserving sign and time values.

Example:

```
(X + Y)[-5 ns]
-- is equivalent to
(X[-5 ns] + Y[-5 ns]).
-- and
(X + 2) [3 ns]
-- is equivalent to
(X[3 ns] + 2)
```

Example:

```
(X[-5 ns] + Y[-2 ns])[3 ns]
-- is equivalent to
(X[-2 ns] + Y[1 ns])
-- because X[-5 ns + 3 ns] =  X[-2 ns],
-- and Y[-2 ns + 3 ns] = Y[1 ns]
-- after distribution of time 3 ns
-- over objects X and Y.
```

Each expression in VAL is regarded as a waveform stretching from the past to the future. So, another interpretation of a timed expression is to regard the timing operator denoted by [ ] as the operator to shift the waveform of the expression forward (i.e., to the future) or backward (i.e., to the past) in time according to the time value. A positive time value shifts the waveform to the future while a negative time value shifts the waveform to the past. The above examples can be reinterpreted by shifting the waveform. Refer to waveform algebra for details about formal definition of timing operations and waveforms [1].

15.2 Intervals

VAL provides the construct called *timing qualifier* that applies to expressions. Application of timing qualifiers is limited to boolean expressions. A qualifier allows to check the stability of an expression over a period of time (**during**) or to relax constraints on the values of an expression (**within**). See Sections 2.1.4.2 and 2.1.4.4 for an informal introduction to time qualifiers.

Boolean expressions with timing qualifiers are called *time qualified boolean expressions*, and have the following syntax:

```
time_qualified_boolean_expression ::=
    simple_qualified_boolean_expression
  { logical_operator
    simple_qualified_boolean_expression }

simple_qualified_boolean_expression ::=
    boolean_expression during time_interval
    boolean_expression within time_interval

time_interval ::=
    '[' time_expression, time_expression ']'
  |     time_expression
```

The time qualified boolean expression consists of a boolean expression, timing qualifier keywords **during** or **within**, and a time interval. The time interval consists of time expressions indicating the time period to evaluate the value of the boolean expression. The time expression should have time units of VHDL physical type `TIME`. VAL time expressions can have both positive and negative values.

The semantics of qualified boolean expressions are defined as follows. A boolean expression qualified by **during** is true only if the pure boolean expression (without the time qualifier) has the value true throughout the given time interval. Otherwise, it is false. A boolean expression qualified by **within** is true only if the pure boolean expression (without the time qualifier) has the value true at least once within the given time interval. Otherwise, it is false.

A time interval is specified by two numeric expressions enclosed by square brackets. The first numeric value indicates the starting time point and the second one indicates the ending time point of the time interval. VAL requires that the ending time value should be always greater than or at least equal to the starting time value. Since time 0 is used as the current time, a negative value means the past and the positive value means the future. Note that time is relative to the waveform of the boolean expression, and thus the current time 0 is nothing but a reference point (see section 2.1.4.3).

When there is only one time expression after the time qualifier without brackets denoting a time interval, the time expression should have a non-negative time value. In such a case, the time interval is a simplified form of interval starting at the negation of the time expression and ending at the current time (i.e., time 0).

Some syntactic restrictions are also required on the occurrence of qualifiers in expressions. The qualifiers **during** and **within** have the lowest precedence in terms of association in expressions. A qualified boolean expression cannot syntactically contain another qualified expression. Parentheses must surround nested qualifiers. A qualified boolean expression can be negated by prefixing the logical operator **not**. For example, the expression **not** (e_1 **during** t_1) means logical negation of the evaluation of the result of expression (e_1 **during** t_1).

Example:

```
X = 5 during [-10 ns, 0 ns]
```

The above expression evaluates to true if `X` has maintained value 5 for the last 10 ns. Otherwise, it evaluates to false.

Example:

```
X = Y + Z during [-15 ms, 5 ns]
```

The above expression will evaluate true if `(X = Y + Z)` has been true for the last 15 ms and will remain true for at least 5 ns in the future from the current time. Otherwise, it evaluates false.

Example:

```
X = 7 during [-5 ns, -5 ns]
```

The above expression evaluates to true if `X` had value 7 at 5 ns in the past; otherwise, it evaluates to false. In fact, the above qualified expression is equivalent to

```
(X = 7) [-5 ns].
```

If the ending point in the time interval is the current time (i.e., time 0), the time interval can be specified by one time value.

Example:

```
X = 5 during 10 ns
```

is equivalent to:

```
X = 5 during [-10 ns, 0 ns]
```

In the above short form of time qualified expression, the time value should be non-negative.

Note: Another semantic interpretation of time qualifier from the waveform standpoint is that the time qualifier reduces the width of truth of the boolean expression over time. Generalization of the VAL time qualifier is discussed in [1].

Restrictions on Time Expressions

Expressions representing time (e.g., expressions inside `[` and `]`) in timed expressions and time qualified boolean expressions should be static expressions. The definition of a static expression follows the one defined by VHDL. (Refer to Section 7.4 of VHDL LRM [25].)

15.3 Function Call

The function call has the syntax of the VHDL function call. It allows to abstract a complex expression under one name. The function call semantics and semantics restriction are the same as the VHDL function call semantics.

Chapter 16

Statements

VAL statements appear in the VHDL entity statement part and in the VHDL architecture statement part.

```
VAL_entity_statement ::=
        assertion_statement
    | guarded_statement
    | select_statement
    | drive_statement
    | generate_statement
    | macro_call_statement
    | null_statement

VAL_architecture_statement ::=
        assertion_statement
    | guarded_statement
    | select_statement
    | generate_statement
    | macro_call_statement
    | null_statement
```

Neither the entity state nor signals can be modified by body annotations in the VHDL architecture. Thus, the drive statement is not allowed in body annotations.

VAL architecture statements are used to describe assertions and constraints on the behavior of a VHDL architecture. Annotations that describe relationships between the entity state declared in the entity annotations and its implementation in the VHDL architecture are called *mapping* annotations, Mapping annotations allow more detailed checking of consistency between entity annotations and the architecture body during simulation. (See Chapter 17 for mapping annotations.)

This chapter describes the assertion statement, guarded statement, select statement, drive statement, generate statement, macro call statement, and null statement.

In VAL, the assertion and drive statements are *processes* (see section 2.1.3). Key to the understanding of VAL is the concept of *guard*. A guard is a boolean condition that controls the execution of some or all statements in its scope (see the VHDL LRM, Sections 9.1, 9.5). VHDL guards appear in block declarations and control only concurrent assignment statements. VAL guards are defined by guarded statements (see section 16.3), and control all statements in their scope. A VAL statement is said to be *active* if all its guards evaluate to true. Guards are defined in guarded and select statements. A process executes when it first becomes *active* or when it is active and there is an event on any of its signals.

Generate and macro call statements are supported as syntactic sugar to simplify description of annotations. Generate statements are used to replicate a statement or a list of statements. A macro call statement instantiates a macro text that is defined as a macro declaration. Macros permit a list of VAL statements to be abstracted into a single macro call.

16.1 Assertions

Assertions express conditions and constraints in terms of ports, signals, variables, constants, generics, and entity states that must be satisfied during VHDL simulation.

An assertion statement consists of a keyword indicating the assertion flavor, a boolean expression, and an optional severity and report specification.

```
assertion_statement ::=
    assertion_flavor boolean_extended_expression
  [ else ]
```

```
        [ report expression ]
        [ severity expression ] ;

assertion_flavor ::=
      assert
    | finally
    | sometime
    | eventually
```

If the **report** clause is present, it must include an expression of predefined type **STRING** that specifies a message to be reported. If the **severity** clause is present, it must specify an expression of the predefined type **SEVERITY_LEVEL** that specifies the severity level of the assertion. **SEVERITY_LEVEL** is an enumeration type consisting of (**NOTE**, **WARNING**, **ERROR**, **FAILURE**) as defined in VHDL.

The **report** clause specifies a message string to be included in the error message generated by the assertion. In the absence of the **report** message, the default value for the message string is **Assertion violation**. The **severity** clause specifies a severity level associated with the assertion. In the absence of a **severity** clause, the default value of the severity level in VAL assertions is **WARNING**. This is in contrast to VHDL which has **ERROR** as a default severity level for assertions.

VAL introduces the optional keyword **else** in the assertion statement for the following reason. Assertion means the condition that has to be satisfied, and report and severity specify the action to be taken in case the condition is not satisfied. By adding the keyword **else**, this situation is distinguished more clearly, and thus improves readability and writability.

The assertion statement defines a process. When it is active, the assertion process evaluates the boolean expression. If the boolean expression evaluates to false, the assertion process reports the message specified in the **report** clause. Severity level **ERROR** forces simulation to stop, while severity level **WARNING** allows the simulation to proceed after the assertion process has reported its message.

VAL augments the VHDL assertion statement by introducing assertion flavors. Assertion flavors enable the user to reduce the number of simulation cycles at which the assertions are checked. The following discussion relies heavily on the concept of simulation cycle (see section 12.6.3 of the VHDL manual for reference). Simulation cycles are also called *deltas*. One characteristic of a delta is its time stamp.

Assuming a delta with time stamp 10ns, and an event occurring on signal Y during that delta, an assignment such as:

```
X <= Y after 3ns.
```

will generate a new delta with time stamp 13ns. With the same assumptions, an assignment such as:

```
X <= Y ;
```

will not change the value of X during the current delta, but will generate a new delta, with the same time stamp. The event on X is said to have happened one delta later than the event on X, hence the term *delta delay*. The set of all simulation cycles at a given simulated time is a *time point*.

In VHDL, assertions are evaluated at every simulation cycle. This may be too strict to apply VAL assertions in annotations for two reasons:

- First, the user often has to count the exact number of deltas resulting from zero delay assignments in using assertions. For a given signal assignment statement `X <= Y;`, it is tempting to use the assertion statement `assert X = Y;`. However, this assertion is not correct because of delta delay associated with the assignment. The correct assertion is `assert X = Y'delay;`. In general, it will be very tedious to count the exact number of deltas in the VHDL descriptions.

- Second, there are cases when it is is impossible to assert some behavior in the VHDL entity declaration that will be used to check different architectures (or implementations) of the entity. The number of deltas may be different according to the architecture. Thus, it is impossible to assert architecture-independent common behavior in the entity declaration, and this situation severely restricts the usefulness of VHDL assertions.

VAL assertion flavors alleviate the above problems and generalize the VHDL assertion capability. Assertion flavors allow assertions to be checked only at meaningful simulation cycles during simulation. They can exclude assertion checking at irrelevant simulation cycles such as the transient period dependent on a particular computation sequence. Assertions capture the various behaviors expected at observable points in time during the course of a simulation. Different assertion flavors help designers to use assertions more easily and naturally.

There are four flavors of assertions in VAL:

`assert` The assertion must be always true while the assertion process is active. Thus, the assertion must be true at every delta during VHDL simulation where the guards of the assertion process are true. (See section 16.3 for a description of guards).

`eventually` When the assertion process is active, the assertion must become true during some simulation cycle at the current simulated time. From then on, the assertion stays true until the assertion process is deactivated.

`finally` The assertion must be true at the end of the time point during which the assertion process becomes active. Thus, the assertion must become true during the last simulation cycle (at the latest) of the time point in which the assertion process becomes active.

`sometime` The assertion must be true at some simulation cycle during the time point in which the assertion process is active.

In the above description, the assertion means a boolean (or extended boolean) expression in the assert statement (see section 2.1.3.1).

When an assertion statement is guarded, assertion checking is done only while the assertion statement is active (i.e., the guard is true).

Example:

The output signal `y` is equal to the input signal `x` delayed by 10 ns at every delta. If the assertion is not satisfied, stop the execution and report the error message.

```
--| assert Y = X[-10 ns] else
--|     report "Output error"
--|     severity error;
```

However, if we are interested in observing the behavior of the signals only at the last simulation cycle of a given time point, the assertion should be specified with the **finally** flavor.

```
--| finally Y = X[-10 ns] else
--|     report "Output error"
--|     severity error;
```

Example:

```
-- Assertion with default message
-- and severity level WARNING
--| assert Y = X[-10 ns];
```

Because of relative time in VAL, the above assertion is equivalent to the following assertions:

```
--| assert Y[10 ns] = X;
```

```
--| assert Y[5 ns] = X[-5 ns];
```

Example:

```
-- Assertions with different flavors
--| assert Z = X and Y;
--| finally Z = X and Y;
--| eventually Z = X and Y;
--| sometime Z = X and Y;
```

16.2 Drive Statement

A drive statement is used to modify the entity state. Since modification of the entity state is allowed only inside the entity declaration, drive statements are allowed only in the entity declaration. It is illegal to use a drive statement in annotations of an architecture body or a configuration specification.

```
drive_statement ::=
    extended_expression -> object_name [ time_indicator ] ;
  | object_name [ time_indicator ] <- extended_expression ;

time_indicator ::= '[' time_expression ']'
```

The target object of a drive statement must be either the entity state (when the entity state is of non-composite type) or a component of an entity state (when it is of a composite type such as array or record). The target can be either on the left or right side of the drive statement according to the driving operator (i.e., <- or ->). The type of the target object must be compatible with the type of the expression.

The optional time indicator associated with the target object in the drive statement must be a static expression whose evaluation has type **TIME**. Type **TIME** of VAL is the same as the predefined physical type **TIME** of VHDL except for the following difference: whereas VHDL requires that the value of **TIME** must be a non-negative number, VAL allows a negative value.

The drive statement defines a process that evaluates an expression and assigns its value to a target object. As long as the drive statement is active, the value associated with the object is changed to match the value of the expression. If at any point the value of the expression changes, the value of the object also immediately changes.

It is an error for the same target object to be driven during the same delta by more than one drive statement. When an entity state is not driven, it keeps the value it was last assigned.

Example:

```
--| not X1 -> STATE.Z1;
--| X2 and  X3 -> STATE.Z2;
--| Y1 + 3 -> STATE.Z3;
```

In the above example, all three drive statements execute in parallel.

For specification of a timing behavior of an entity state, timed expressions and timed objects are used in the drive statement.

Example:

```
--|  (X + Y)  -> STATE[10ns];
```

The above example describes that the sum of the current values of X and Y drives the entity state STATE at 10 nsec in the future. Actually, 10 nsec represents the time delay from an add operation to the actual assignment of the operation result to STATE.

Since VAL deals with relative time, the above drive statement is equivalent to each of the following statements:

```
--|  (X + Y)[-10ns]  -> STATE;
--|  (X + Y)[-5ns]  -> STATE[5ns];
--|  (X + Y)[-15ns] -> STATE[-5ns];
```

Default delay of a drive statement:

When a driving action involves a delay in modeling a hardware behavior, the time is specified using a timing construct in a drive statement. However, when the delay omitted or zero delay is specified, the VHDL delta delay semantics apply.

Anticipatory semantics:

Since VAL only monitors the VHDL simulation with assertions, a drive statement in VAL has anticipatory semantics [16, 17]. Anticipatory semantics mean that once a drive statement assigns a future value to an object, the object will always have the value when time proceeds to the future time. Anticipatory semantics are in contrast to preemptive semantics used in VHDL's assignment statements. VHDL's preemptive semantics mean that projected transactions may be eliminated from the transaction queue if they are found to be incorrect projections. The VHDL inertial and transport delays are defined based on preemption of transactions.

Causality principle:

When an entity state or its component is driven by some value, it must satisfy the causality principle. This means that a past value cannot be changed by the current or future value. In the drive statement `X[t1] -> Y[t2]`, time t2 must be greater than or equal to time t1. For example, `X[5] -> Y[3]` is illegal because it is not causal (i.e., the future value of `X` is driving the past value of `Y`.)

Multiple driving:

The target object must be driven by at most one drive statement at a time. Thus, it is a semantic error if the same object is driven at the same point in time by more than one drive statement. The VAL runtime system will check the multiple driving to the entity state during simulation, and report an error if a violation is detected.

Example:

When the following statements are in the entity annotations, it is an error:

```
--| not X1 -> STATE;
--| X2 and X3 -> STATE;
```

When an entity state has a composite type, the granularity of concurrent driving is the component level of the type. Thus, the following statements are legal in the entity annotations:

```
--| not X1 -> STATE.Z1;
--| X2 and X3 -> STATE.Z2;
```

Example:

```
--| when P then
--|     STATE <- X;
--| end when;
--| when Q then
--|     STATE <- Y;
--| end when;
```

In the above example, if there is any case where both `P` and `Q` are true at the same time during simulation, it is reported as a multiple driving error.

16.3 Guards

A guarded statement is a basic control structure in VAL. Guards control activation and deactivation of a list of statements. A guarded statement consists of a boolean expression called guard, a list of statements belonging to its **then** part, and an optional list of statements belonging to its **elsewhen** or **else** parts. Guards can be nested hierarchically.

```
guarded_statement ::=
    when boolean_extended_expression then
        VAL_statement_list
  { elsewhen boolean_extended_expression then
        VAL_statement_list }
  [ else
        VAL_statement_list ]
    end when ;
```

When active, a guarded statement checks the value of the guard expression. If the expression is true, the list of statements belonging to its **then** part are activated. Otherwise, the list of statements belonging to its **else** part are activated. If there are any **elsewhen** parts, when the guard expression after **when** evaluates to FALSE, guard expressions after **elsewhen** are evaluated in sequence either until one evaluates to TRUE, or the **else** has been reached. Then, the list of statements in the corresponding part will be activated. If the value of the guard expression changes, the statements corresponding to the newly activated branch of the guarded statement become active and those corresponding to the other branch are immediately deactivated. Thus a guarded statement represents one node of a tree describing the branches of control in VAL annotations. A guarded statement, like any other statement at the top level of the behavioral specification, is always active. By definition, a guarded statement at any other level must be nested within one or more guarded statements, called its parents. A guarded statement at any other level is active if and only if it is activated by its immediate parent.

The guarded statement must also observe the causality principle. At the moment of performing an assignment to the entity state, all the decisions expressed by the guard(s) must be made based on current and past information.

The way to check for causality is to choose a reference point (*rescale*, see section 2.1.4.3) such that the target of the guarded drive statement has a 0ns time qualifier (if there is more than one guarded drive statement, this may require different time shifts for each statement). The process of rescaling until the target of the drive statement as a 0ns time qualifier is called *normalization*. All guards must then refer to values of signals in the past.

The following guarded statement:

```
--| when Y[3 ns] then
--|     STATE <- X[-2 ns];
--| end when;
```

is illegal because it violates causality. It reads as follows: if **Y** is true 3 ns in the future from now, **STATE** gets the value of **X** 2 ns ago. Thus, driving to the current **STATE** is determined by the guard using future information.

In order to observe causality, the timing in the guarded statement must satisfy the following restriction. For a given guarded statement obtained after normalization through rescaling of time,

```
--| when X[t1] then
--|     STATE <- X[t2];
--| end when;
```

time **t1** and **t2** should be less than or equal to zero in order for the statement to be causal.

Example:

At each delta, whenever port **X** goes high, check if it has been low for at least 5 msec.

```
--| when X'changed('1') then
--|      assert X = '0' during [-5ms, 0ms];
--| end when;
```

X'changed is the predefined VAL attribute associated with signals and entity states.

Example:

Check for each delta that the input signal **X** is stable during a setup and hold time around the rising edge of the clock.

```
--| when clock'changed('1') then
--|     assert x'Stable during [-setup, hold]
--|          report "Setup error";
--| end when;
```

Example:

While **select_port** has value 1, **STATE** is driven by the value of **X**; otherwise, **STATE** is driven by the negated value of **X**.

```
--| when select_port = 1 then
--|      X -> STATE;
--| else
--|      not X -> STATE;
--| end when;
```

Guards can be nested. When a drive statement has nested guards, it is active only when all the nested guards are true; otherwise, it is inactive.

Example:

```
--| when select_port = 1 then
--|     assert STATE.x1 = Z1;
--|     when clock = 1 then
--|         assert STATE.x2 = Z2;
--|     end when;
--| end when;
```

In the above example, **Z1** is asserted to be equal to the value of **STATE.x1** while **select_port = 1** is true, and **Z2** is asserted to be equal to the current value of **STATE.x2** only when both **select_port = 1** and **clock = 1** are true.

Example:

This example shows three equivalent VAL descriptions for the following behavior: If the input signal **x** is stable during a setup and hold time around the rising edge of the clock, change the value of the internal state **y** to the value on the input signal.

If we take the rising edge of clock (i.e., **clock'changed('1')**) as the reference time, the VAL description is:

```
--| when clock'changed('1') then
--|     when x'stable during [-setup, hold] then
--|         x -> y[delay];
--|     end when;
--| end when;
```

After normalization of time, the VAL description is:

```
--| when clock'changed('1')[-delay] then
--|     when x'stable during [-setup - delay,
--|                             hold - delay] then
--|         x[-delay] -> y[0ns];
--|     end when;
--| end when;
```

When nested guards are flattened, the VAL description is:

```
--| when (clock'changed('1')) and
```

```
--|        (x'stable during [-setup, hold]) then
--|      x -> y[delay];
--| end when;
```

In the above example, note that **delay >= hold** must be true to satisfy causality.

Example:

```
--| when select_port = 1 then
--|      finally outport = input1 ;
--| elsewhen select_port = 2 then
--|      finally outport = input2 ;
--| else
--|      finally outport = input3 ;
--| end when;
```

is equivalent to:

```
--| when select_port = 1 then
--|      finally outport = input1;
--| end when;
--| when not(select_port = 1) and
--|           (select_port = 2) then
--|      finally outport = input2 ;
--| end when;
--| when not(select_port = 1) and
--|       not (select_port = 2) then
--|      finally outport = input3 ;
--| end when;
```

16.4 Select

A select statement chooses the activation of one of a number of alternative lists of statements; the chosen alternative is defined by the value of an expression. A select statement provides collective guards in a restricted form where guards are mutually exclusive. In other words, the select statement is a shorthand for several guarded statements with mutually exclusive guards.

```
select_statement ::=
    select extended_expression is
        select_statement_alternative
      { select_statement_alternative }
    end select ;

select_statement_alternative ::=
    in choice
  { '|' choice } => VAL_statement_list
choice ::=
    simple_expression
  | discrete_range
  | component_simple_name
  | others
```

In the above description, '|' means the actual | character in the VHDL text, as opposed to the BNF metacharacter |. The expression after **select** must be of a discrete type, and each choice in a select statement alternative must be of the same type as the expression. In the last alternative, the choice can be **others**. An **in others** alternative covers all cases that previous alternatives did not cover.

The choices must be mutually exclusive and the collection of all choices must cover the whole range of the discrete type for the selection.

A select statement is semantically equivalent to a set of mutually exclusive guarded statements. It evaluates an expression and executes one of several lists of statements, based on the value of the expression. Each list of statements is prefixed by a list of values called the choice list. When the select expression matches one of the values in the choice list, the list of statements associated with that choice is activated. The expression must evaluate to the value of a discrete type. If none of the choices matches the expression, the statements in the **in others** alternative are activated.

Each choice list must be distinct and the compiler must be able to verify this at compile time. Thus the compiler must be able to evaluate expressions within a choice list at compile time. For this purpose, extended expression and choices must be all locally static.

Example:

```
--| select auto_diagnosis_result is
--|     in worn_out_break_pad => replace_break_pad ;
--|     in no_oil => fill_the_oil ;
--|     in others => see_a_mechanic ;
--| end select;
```

Example:

```
--| select traffic_state(0 to 1) is
--|     in '00' =>
--|         finally Highway_Light = green;
--|         finally FarmRoad_Light = red;
--|     in '01' =>
--|         finally Highway_Light = yellow;
--|         finally FarmRoad_Light = red;
--|     in '10' =>
--|         finally Highway_Light = red;
--|         finally FarmRoad_Light = green;
--|     in '11' =>
--|         finally Highway_Light= red;
--|         finally FarmRoad_Light = yellow;
--| end select;
```

Example:

```
--| select op_code(0 to 1) is
--|     in '00' => NOOP;
--|     in '01' => BRANCH;
--|     in '10' => SHIFT;
--|     in '11' => ADD;
--| end select;
```

16.5 Generate

A generate statement replicates a statement or a list of statements according to indexed values. Note that generated statements are all executed in parallel. When a family of statements are parameterized by index values, they can be represented in a simpler form by using a generate statement.

```
generate_statement ::=
    VAL_generation_scheme generate
        VAL_statement_list
    end generate ;
VAL_generation_scheme ::=
    for generate_parameter_specification
parameter_specification ::= ... (Same as VHDL)
```

The syntax of a generate statement follows VHDL. The generate parameter specification is the declaration of the *generate parameter* with the given identifier. The generate parameter is a constant object whose type is the base type of the discrete range of the generate parameter specification.

A generate statement is nothing but syntactic sugar to allow a family of statements having array type names to be described in a simple form by parameterizing array indices.

Example:

The generate statement

```
--| for i in 0..3 generate
--|     Y(i) -> Y(i+1);
--| end generate ;
```

is equivalent to the following set of parallel statements:

```
--| X(0) -> Y(1) ;
--| X(1) -> Y(2) ;
--| X(2) -> Y(3) ;
--| X(3) -> Y(4) ;
```

Example:

Generate statements are nested when there is more than one parameter:

```
--| for i in 0 to 1 generate
--|     for j in 2 to 3 generate
--|         finally X(i) = Z(j);
--|     end generate;
--| end generate;
```

16.6 Macro Call

A macro is used in abstracting a common behavior that appears at multiple places in annotations. A macro call statement instantiates a macro that is declared previously in the VAL declarative part or in a separate package. The instance of a macro is expanded textually using the corresponding macro body during the VAL preprocessing.

```
macro_call_statement ::=
    macro_name ( association_list ) ;

association_list ::=
    association_element { , association_element }
association_element ::= [ formal_part => ] actual_part
formal_part ::= identifier
actual_part ::= expression
```

When a macro is declared with parameters, a macro call statement has to provide arguments that match the macro parameters. Arguments may be designated by position or by name. Arguments for all parameters must be specified; default arguments for parameters are not allowed. Parameters are substituted by corresponding arguments during macro expansion. It is prohibited to call a macro recursively. Thus, it is illegal to call a macro with the same macro name as an actual parameter.

Example:
If we assume the following macro declaration:

```
--| macro check_state (condition, message, sev_level) is
--|     assert STATE.status = condition
--|         report message
```

```
--|          severity sev_level;
--| end check_state;
```

the macro call

```
--|  check_state (safe_state, "Fatal error", ERROR);
```

will be expanded as follows:

```
--| assert STATE.status = safe_state
--|     report "Fatal error"
--|     severity ERROR;
```

Example:

Since a macro may not be called recursively, the following macro call is illegal:

```
--| check_state (check_state, "Fatal error", ERROR);
```

16.7 Null

A null statement is a statement performing no action.

```
null_statement ::= null ;
```

The execution of the null statement has no effect on the environment.

Chapter 17

Mapping Annotations

Mapping annotations are used to specify how the VAL entity state declared in the entity annotations is implemented in the architecture body. The VHDL implementation of the entity state is made of objects declared in the VHDL architecture. Those objects are signals declared in the architecture and its blocks, as well as the entity ports, the ports and state of all architecture components. From the VAL standpoint, variables declared inside processes are not treated as implementation objects.

Mapping annotations are VAL assertions in the architecture body that express relationships between the VAL entity state and its implementation in terms of architecture objects. Thus, hardware designers can specify relationships between the VAL entity specification and the architecture body.

Mapping annotations are used to check behavioral consistency between the VAL entity specification and the VHDL architecture body. The consistency checking is obtained automatically by evaluating assertions during the course of simulation. When any inconsistency occurs, it will be detected by a violation of VAL mapping assertions reported during the simulation. The messages reported from assertion violations will help the user to locate the source of the inconsistency. Consistency in a multi-level design hierarchy is achieved by repetitive application of consistency checking between two adjacent levels (i.e., VAL entity annotation and the VHDL architecture body) across the design hierarchy.

In order to facilitate mapping annotations, VAL provides the following: (1) visibility of entity state in an architecture, and (2) visibility of the state of local components in an architecture.

Visibility of entity state

The VAL entity state declared in the entity annotations is also visible to annotations in the architecture body of the same entity. The visibility rule of the entity state is similar to that of ports. The difference is that the entity state is visible only to annotations in the architecture whereas ports are visible to both VHDL source code and VAL annotations in the architecture.

Architecture annotations can read both the entity state and ports declared in the entity declaration. However, architecture annotations cannot modify the value of the entity state or ports.

Visibility of component state

Each component declared in the architecture body also has its own entity state, called a component state. VAL supports annotations to declare the *local state model* declaration in order to provide the visibility of component state in the architecture body.

The local state model declaration is declared as an annotation inside the VHDL component declaration in the architecture body. The syntax for the local state model declaration is the same as the entity state model declaration.

```
annotated_component_declaration ::=
    component identifier
        [ local_generic_clause ]
        [ local_port_clause ]
        [ --| local_state_model_annotation ]
    end component ;

local_state_model_annotation ::=
    state_model_declaration
```

By means of the local state model declaration, the entity state of a component becomes visible in the architecture body containing the component. (Refer to Section 4.5 of VHDL LRM [25] for component declaration.)

Architecture annotations can access component states on a read-only basis; they cannot modify entity states of components. These component states can be modified only by VAL entity annotations in the entity declaration of the corresponding component.

The entity state of a component is referred to by the state object name after the component name. For example, assuming the default name **state** is used for the entity state, the state of a component instantiated with the name **X** is referred to by **X.STATE**. If an entity state has name **Y**, the component state of the instantiated component **X** is denoted by **X.Y**.

Example:

```
architecture structure of CPU is
...
  component shift_register
    port (A : in BIT; B : out BIT);
    -- local state model of register
    --| state model is BIT_VECTOR (0 to 15) := initial_value;
  end component;
...
begin
  -- Instantiate register as a component whose name is D1
  D1: shift_register
     port map (A => S1, B => S2);
  ...
  -- Mapping assertion to relate
  -- component "reg1" of the entity
  -- state of CPU and the entity
  -- state of component D1.
  --| finally STATE.reg1 = D1.STATE;
  ...
end structure;
```

A component declaration defines an assumed design entity that will be instantiated as a component in the VHDL architecture. The local state model declaration in the component declaration assumes a state model of the component that is referred to by annotations in the architecture body. This locally assumed entity state of a component will be associated with the actual state model declared in the entity annotation. The binding of the component declared locally in architecture with the actual entity declaration is specified by the VHDL component configuration or the VHDL configuration specification. If the local state model of a component is different from the state model of the actual entity, it has to be resolved by the VAL state model map annotation. This state model map annotation declares the type conversion function for two different state models. The state model map annotation is declared in the annotation region of the configuration specification described in Chapter 18.

Chapter 18

Configuration Annotations

Annotations in the configuration specification are used to specify (1) inclusion of VAL annotated design entities such as annotated entity declarations and architectures for simulation, and (2) type conversion for different state models.

The primary use of configuration annotations is to specify the binding of instances of components to VAL annotated design entities. In a hierarchical design, VHDL configuration declarations and configuration specifications are used to specify how each instance of a component is bound to the VHDL architecture body of the component at the lower level. These VHDL configuration specifications determine the design hierarchy that is represented as a tree whose leaves consist of only purely behavioral architecture bodies. The VHDL simulation will be executed based on this configuration.

For simulation purposes, VAL annotations are translated as monitoring VHDL entities, or *monitors.* The user can specify which monitors to include in the simulation using configuration annotations. For example, assuming a CPU made of a data path and a control unit, the user can decide to monitor the data path but not the control unit. For the purpose of efficiency in simulation, the user may decide not to monitor some components of his design.

Another type of configuration annotation is the *state model map*, used for type conversion between different state models. When a component declaration in the architecture body has an annotation for a local state model declaration, the local state model may be different from the state model declared in the actual entity declaration. This situation often occurs in a hierarchical design or multi-designer environment. In top-down design, a user can assume a component state model inside the architecture if the component is not yet defined. When the component is defined later as an entity declaration and an architecture, it may have a different type for the state model. For example, the component state model is assumed to be an **integer** in the component declaration while the entity state model of the actual entity for the component is declared as type **bit_vector**. In bottom-up design, the same problem can occur.

VAL allows a component state model to be different from the state model of the actual entity. The difference between two state models is resolved by using the state model map facility of VAL. The state model map is declared as an annotation at the place where the configuration for the component is declared (i.e., either a VHDL component configuration that appears in the architecture body or a VHDL configuration specification that exists separately).

Configuration Annotations in Configuration Declarations

Configuration annotations appear in VHDL configuration declarations that are separate design units.

```
Annotated_configuration_declaration ::=
    configuration identifier of entity_name is
       --| VAL_entity_binding_specification ;
        configuration_declarative_part
        block_configuration
    end [ configuration_simple_name ] ;

VAL_entity_binding_specification ::=
  [ valentity ] [ valarchitecture ]

configuration_declarative_part ::= ...(same as VHDL)

block_configuration ::=
    for block_specification
       { use_clause }
       { Annotated_configuration_item }
    end for ;
```

```
Annotated_configuration_item ::=
    block_configuration
  | component_configuration
  | Annotated_component_configuration

block_configuration ::= ...(same as VHDL)
component_configuration ::= ...(same as VHDL)

Annotated_component_configuration ::=
    for component_specification
      [ use binding_indication ; ]
      [ --| state_model_map_annotation ; ]
      [ --| VAL_binding_specification ; ]
      [ block_configuration ]
    end for ;

component_specification ::= ...(same as VHDL)
    binding_indication ::= ...(same as VHDL)

state_model_map_annotation ::=
    state model map (type_conversion_function_name)
VAL_binding_specification ::=
    VAL_entity_binding_specification
  | valconfiguration
```

The type conversion function name in the state model map annotation indicates the name of the function to convert the type of the local state model declared in the architecture to that of the state model of the entity. Conversion functions follow the syntax and semantics of VHDL conversion functions and are declared in a separate package.

Annotations for the state model map appearing in the VHDL configuration declaration should appear only inside the component specification.

Annotations for a VAL binding specification in the VHDL configuration declaration have the following syntactic and semantic restrictions:

- The state model map annotation must appear inside VHDL component specifications.

- The VAL entity binding specifications may appear right at the beginning of the VHDL configuration declaration as well as inside the component specification. In the following example,

```
configuration X of Y is
    --| valentity;
    --| valarchitecture;
    for Z
        -- Component configuration
        -- follows
        ...
    end for;
end X;
```

valentity indicates that the annotated entity declaration for Y should be included in the configuration and valarchitecture indicates that annotations in architecture Z should also be included in the configuration for simulation.

- The use of keywords representing VAL binding specification is restricted depending on the VHDL binding indication. If the VHDL binding indication specifies a VHDL entity declaration and its architecture, the VAL binding specification is restricted to the keywords valentity and/or valarchitecture. In the following example,

```
configuration X of Y is
    --| valentity;
    --| valarchitecture;
    for Z
        for C1: C_type1
            use entity E1(A1);
            --| valentity;
            --| valarchitecture;
        end for;
        ...
        -- other component specifications
    end for ;
end X;
```

the annotation of valentity in the component configuration for C1 indicates inclusion of VAL entity annotations for entity E1 in the configuration, and the keyword valarchitecture indicates that annotations in the architecture A1 of entity E1 should be included in the configuration.

- When the VHDL binding indication specifies another configuration instead of any VHDL entity and architecture, the VAL binding specification is restricted to the keyword valconfiguration. In the following example,

```
configuration X1 of Y is
    --| valentity;
    --| valarchitecture;
    for Z
        for C1: C_type1
            use configuration X2;
            --| valconfiguration;
        end for;
        ...
        -- other component specifications
    end for;
end X;
```

the keyword valconfiguration indicates that the annotated configuration of X2 should be used in the configuration for component C1. In other words, if configuration X2 has any configuration annotations, all those annotations are included for C1. If keyword valconfiguration is missing in the above example, configuration X2 without any annotated units will be used in the configuration (i.e., no annotations will be included).

- the use of the valarchitecture is not always allowed. For example, for an entity X with an entity state model declaration, with an architecture that has mapping annotations and whose components C1, C2, ... have state model declarations, it does not make sense to specify a valarchitecture configuration annotation without specifying a valentity configuration annotation for both the entity X and all its components. Indeed, the mapping annotations need the state model to be properly maintained (which happens only if the valentity keyword is specified), in order to provide meaningful monitoring results. The rule is as follows: if the user wishes to specify a **valarchitecture** configuration annotation for an architecture **A**, he must also specify a **valentity** configuration annotation for all entities whose state is referred to in the architecture **A** annotations.

Configuration Annotations in VHDL Architectures

Configuration annotations can also appear in the declarative region of the VHDL architecture body when the architecture has configuration specifications. The syntax is:

```
conf_state_model_map_annotation ::=
    for component_specification
                state_model_map_annotation
conf_VAL_binding_specification ::=
    for component_specification
                VAL_binding_specification
```

Because the VHDL syntax for component declaration is in the form of a simple statement without the closing keyword **end**, annotations declaring the state model map and the VAL binding specification in the architecture body appear as separate declarative items. These VAL declarations should appear immediately following the declaration of the corresponding configuration specification.

Example:

```
-- An architecture of a CPU :
architecture structure_view of CPU is
    ...
    component shift_register
        port (A : in BIT; B : out BIT);
        -- local state model of register
        --| state model is array (0 to 15) of BOOLEAN;
    end component;
    ...
begin
    -- Instantiate register as a component whose name is D1
    D1: shift_register
        port map (A => S1, B => S2);
```

```
    ...
end structure_view;
```

```
-- A configuration of the CPU:
-- Cell_Lib contains entity and architecture declarations of
-- components such as SR_Cell, and Convert_Package contains
-- declarations of functions to define type conversions.
library Cell_Lib, Convert_Package;
configuration CMOS_Version of CPU is
  ...
  for structure_view
    for D1: shift_register
      use entity Cell_Lib.SR_Cell(CMOS)
      port map ( ... );
      --| state model map
            (Convert_Package.BoolArray_To_BitArray);
      --  BoolArray_To_BitArray is
      --  the function declared
      --  in Convert_Package that
      --  convert the state models.
      --| valentity;
      --| valarchitecture;
      --  Both entity annotations and
      --  body annotations of shift_register
      --  will be included in the VHDL simulation.
    end for;
  end for ;
  ...
end CMOS_Version;
```

The configuration annotation for component `D1` in the configuration declaration of the above example can be also specified in the architecture body of entity CPU when the configuration is specified in the architecture.

Example:

```
-- An architecture of a CPU
-- with configuration specification:
library Cell_Lib, Convert_Package;
architecture structure_view of CPU is
```

```
    ...
    component shift_register
        port (A : in BIT; B : out BIT);
        -- local state model of register
        --| state model is array (0 to 15) of BOOLEAN;
    end component;
    ...
    -- component configuration
    for all: shift_register
        use entity Cell_Lib.SR_Cell(CMOS);
    -- configuration annotation
    -- for state model mapping
    --| for all : shift_register
    --|    state model map
    --|    (Convert_Package.BoolArray_To_BitArray);
    --  VAL binding specification saying
    --  that both entity annotations and
    --  body annotations of shift_register
    --  will be included in VHDL simulation.
    --| for all : shift_register valentity;
    --| for all : shift_register valarchitecture;
    ...
begin
    ...
    -- Instantiate register as a
    -- component whose name is D1
    D1: shift_register
        port map (A => S1, B => S2);
    ...
end structure_view;
```

Chapter 19

Miscellaneous

19.1 Package

VAL declarations can appear in VHDL packages intermixed with VHDL declarations. As with all VAL annotations, VAL declarations appearing in packages are preceded by --|.

VAL declarations follow the linear visibility rules of VHDL. A VHDL declaration is visible to both VHDL source code and VAL annotations that follow the declaration. In contrast, a VAL declaration is visible only to VAL annotations that follow.

```
VAL_package_declaration ::=
    package identifier is
        VAL_VHDL_package_declarative_part
    end [ package_simple_name ] ;

VAL_VHDL_package_declarative_part ::=
  { VAL_VHDL_package_declarative_item }

VAL_VHDL_package_declarative_item ::=
    --| VAL_package_declarative_item
  | package_declarative_item

package_declarative_item ::= ...(same as VHDL)

VAL_package_declarative_item ::=
```

```
    subprogram_specification
  | type_declaration
  | subtype_declaration
  | constant_declaration
  | alias_declaration
  | use_clause
  | macro_specification
```

VAL deferred constants, function bodies, and macro declarations may appear in VHDL package bodies. When a macro or subprogram specification appears in a package, the corresponding macro or subprogram body (declaration) appears in the associated package body.

```
VAL_package_body_declaration ::=
    package body package_simple_name is
        VAL_VHDL_package_body_declarative_part
    end [ package_simple_name ] ;

VAL_VHDL_package_body_declarative_part ::=
   VAL_VHDL_package_body_declarative_item

VAL_VHDL_package_body_declarative_item ::=
    --| VAL_package_body_declarative_item
  | package_body_declarative_item

package_body_declarative_item ::= ...(same as VHDL)

VAL_package_body_declarative_item ::=
    subprogram_declaration
  | subprogram_body
  | type_declaration
  | subtype_declaration
  | constant_declaration
  | alias_declaration
  | use_clause
  | macro_declaration
```

Example:

```
-- Package declaration
```

```
package Support is
    type Tri is ('0','1','Z');
    function BitVal ( Value : Tri) return BIT ;
        ...
    --| macro Tri_Assert(X, Y) ;
        ...
end Support ;

  -- Package body
package body Support is
    -- body of BitVal
    function BitVal (Value : Tri ) return BIT is
    begin
        ...
    end BitVal ;

    -- body of Tri_Assert
    --| macro Tri_Assert (X, Y) is
    --|       ...{{macro body is here}}
    --| end Tri_Assert ;

end Support ;
```

19.2 Scope and Visibility

This section describes the scope rules for VAL declarations, the visibility rules for identifiers in the annotation text, and the scoping and visibility between VAL annotation text and VHDL text.

19.2.1 Declarative Region and Scope of Declarations

In addition to VHDL declarative regions (see Section 10.1 of VHDL LRM [25]), VAL introduces the following additional declarative regions:

- Macro declarations
- Generate statements

The scoping rules of VAL follow VHDL (see Section 10.1 and 10.2 of the VHDL LRM [25] for scope of declarations) with the following additions:

- The scope rule for annotations are the same as VHDL text. For example, the scope of a VAL declaration is the same as if the preceding annotation indicator --| were removed and the declaration considered to be in the VHDL text at the same point.

- The scope of the index variable declared in the generate statement extends over the region of applicability of the generate statement.

- The scope of formal parameters in a macro declaration is within the macro body.

19.2.2 Visibility

The visibility rules of VAL follow those of VHDL (see Section 10.3 of the VHDL LRM [25]) with the following additions:

- VAL declarations are visible only to annotations in their scope; they are not visible to the VHDL text in the same scope. Thus, VAL declarations cannot hide any VHDL declaration in the same declarative region.

- The entity state declared in the entity annotations is visible to both entity and body annotations that appear after the declaration with this restriction; the entity state can be both read and driven (e.g., modified) by annotations in the entity, but annotations in an architecture body of the entity can only read the entity state.

- The entity state of components in the architecture body is visible in annotations in the architecture body by declaring a local state model in the component declaration. The visibility is restricted to read-only access. The scope is within the block statement containing the component declaration.

- In VHDL, output ports declared in the entity declaration can be assigned but cannot be read from the inside of the entity (i.e., inside of both VHDL entity declaration and architecture bodies). In contrast to VHDL, VAL allows output ports of an entity to be accessible on a read-only basis by both entity and body annotations (i.e., annotations in the entity declaration and the architecture body of the entity) with the following semantic meaning: When a output port is read, it returns the *contribution* of the output port. (See Section 8.3.1 of the VHDL LRM [25] for the meaning of the contribution.)

- The index variable in the generate statement is visible only within the statement after declaration of the index variable. It also hides any other variable with the same name in the statement.
- The visibility rules for formal parameters of a macro declaration and macro specification follow the VHDL rules for subprogram specification and subprogram declaration. (See Section 10.3 of the VHDL LRM [25].)

19.2.3 Use Clause

Syntax and semantics of use clauses in VAL follow those in VHDL. (See Section 10.4 of the VHDL LRM [25].)

19.2.4 The Context of Overload Resolution

VAL supports overload resolution as defined in VHDL. (See Section 10.5 of the VHDL LRM [25].)

19.3 Attributes

VAL supports VHDL predefined attributes of objects and types. In addition, VAL provides the following object attribute:

`S'CHANGED` Returns true if the object `S` is changed at the current point in time; otherwise, it returns false. This is particularly useful to describe behavior based on edges of signal transition.

`S'CHANGED(value)` Returns true if the object `S` changed to the value `value` at the current point in time;

otherwise, it returns false. This is also particularly useful to describe behavior based on edges of signal transition.

Part IV

Transformer Implementation Guide

Chapter 20

The VAL Transformer

In order for VAL annotations to be machine processable, they have to be transformed into some machine executable form. For simulation purposes, there are two possibilities: (1) to transform VAL annotated VHDL into pure VHDL and then use the existing VHDL compiler to transform the pure VHDL into data processable by the simulation engine, or (2) to directly transform the VAL annotated VHDL into data processable by the simulation engine.

This book presents an implementation of the first solution. The advantage of translating VAL to VHDL is that it supports VAL in the context of any VHDL simulator. A VAL compiler would work only in conjunction with one VHDL simulation engine because there is, as of yet, no standard internal representation for VHDL.[1]

The disadvantage of the chosen solution is that it is slower since it transforms VAL into VHDL which is in turn transformed into simulator processable data. Also, being independent of any simulation environment makes it more difficult to use. The user must keep track of the files that are generated by the transformer and put them manually into the design library of the available simulation environment.

[1] An effort is underway, however, to develop a standard internal form.

20.1 Transformation Principles

The translation from VAL to VHDL is based on the following principles supported by the VAL language:

1. *Principle of Separate Compilation.* The transformation of VAL to VHDL is performed on a per compilation unit basis. Each VAL annotated VHDL compilation unit can be transformed into pure VHDL independently from one another. Note that all semantic analysis is performed after the translation by the VHDL Analyzer.

2. *Principle of Name-Transparency.* The VAL to VHDL translation should be hidden from the VAL writer. In other words, the writer of VAL annotations should not need to know about the details of the VAL to VHDL translation. VAL refers only to library names (entity, architecture, and configuration names) visible in the corresponding VHDL context.

   In order to support this principle, the VAL/VHDL configuration declaration contains VAL declarations which state whether a given compilation unit should be used in the original VHDL form or in the translated form. This way the VAL to VHDL translator has information regarding the actual format of a referenced compilation unit; original VHDL format or translated format.

20.2 Translation Methodology

The translation methodology used consists of using a parser generator to generate a parser for VAL annotated VHDL. The parser is extended with a back end that performs the transformation into pure VHDL. The back end uses an abstract syntax tree package in order to access the parse tree, and a symbol table package in order to keep track of all declarations.

The transformation consists of consecutive tree to tree transformations that eventually lead to the desired VHDL internal representation. Each transformation step can be verified using a VAL/VHDL pretty printer that generates the ASCII VAL/VHDL corresponding to a given tree. Note that the pretty printer can print text even if the tree does not correspond to a valid VAL/VHDL source program. This allows the pretty printer to be used for the verification of intermediate transformation steps that leave the tree in an illegal state from a VAL/VHDL source language standpoint.

20.3 Transformation Algorithm

The VAL Transformer runs as a preprocessor on a VAL/VHDL description to generate a self-checking VHDL description. The annotated VHDL description is parsed into a tree format. The transformation algorithm consists of consecutive tree to tree transformations.

In the transformation process, the Transformer must keep track of all VAL and VHDL declarations. This is done using a symbol table.

VAL can annotate three kinds of compilation units: entities, architectures, and configurations. For each compilation unit, there are different kinds of VHDL compilation units being generated. Otherwise, the transformation paradigms for VAL entity annotations and VAL architecture annotations are similar. Some transformations are independent of VHDL (i.e., in terms of VAL only), and others involve VHDL.

As a result of these characteristics of the transformation algorithm, the VAL to VHDL transformation will be presented in five parts:

1. Generation of the transformation skeleton. The VHDL compilation units necessary for each annotated design unit are generated. (See Section 20.3.1.)

2. Transformation of VAL entity annotations into core VAL. Some VAL constructs are re-written in terms of simpler VAL constructs. (See Section 20.3.2.)

3. Transformation of core VAL entity annotations into VHDL. The core subset of VAL is translated into VHDL. (See Section 20.3.3.)

4. Transformation of VAL architecture annotations. Annotations appearing in architectures are translated. (See Section 20.3.4.)

5. Transformation of VAL configuration annotations. Annotations appearing in configurations are translated. (See Section 20.3.5.)

20.3.1 Generation of Translation Skeleton

The translation skeleton is designed to implement the scoping and visibility rules of VAL. Consider the problem of observing the operation of a chip on a circuit board. One way of monitoring the chip is to remove it from its socket, plug a specially constructed adapter into the socket, and then plug the chip into the adapter. The adapter senses the signals traveling between the circuit board and the pins of the chip. The signals can then be monitored to verify the behavior (and use) of the chip.

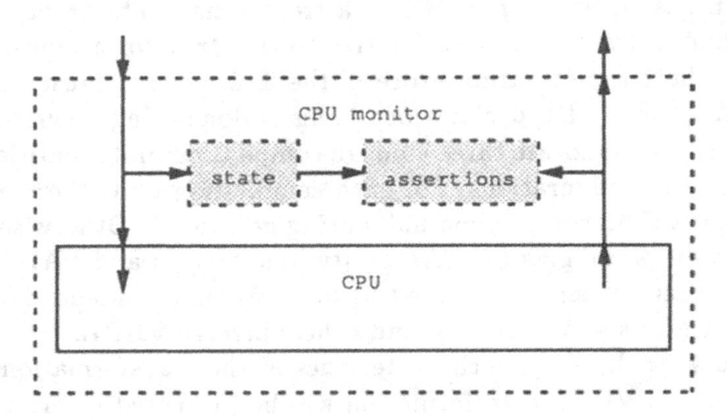

Figure 20.1: Schema of Entity Annotation Translation

For simple cases, the VAL translation algorithm works in just such a manner. Figure 20.1 shows how a new entity is "wrapped" around the entity to be monitored. The Transformer generates an additional architecture called the *Monitor* that contains an instantiation of the component (architecture) under test. The Monitor has the same pins [2] as the component, and contains a *socket* for the component. The Monitor is plugged into a circuit in place of the component. The component is in turn plugged into the socket in the Monitor. In Figure 20.1 the original entity **cpu** is wrapped in an entity `cpu_monitor`.

The Monitor architecture has visibility over all signals traveling be-

[2] Almost. It may have an additional output pin, as described later, to allow other assertions to probe the monitored architecture's internal state.

tween the actual architecture and the other components in the simulation. In addition, the Monitor contains logic to verify the VAL assertions made about the component under test. This includes maintaining its own separate state (the VAL state model).

One advantage of this approach (as opposed to simply monitoring the signals that the pins are connected to) is that VAL assertions can separate the value on out ports of the component from the value on the signals that those ports drive [3]. This allows the user to make assertions about the value placed on the port by the entity. [4]

Consider now an architecture containing several components. If a component is annotated, then a monitor can be generated for that component. The mapping annotations in the architecture have visibility over the internal state of the monitor of the component. This allows annotations within the architecture that "map" the architecture's state into the states of its components. The needed visibility over the internal state of the component is provided through an additional out port on the component that carries the component's state.

Figure 20.2 shows how the translation is structured for a design cpu containing two pla's. The state from the cpu monitor is visible to the architecture annotations in the cpu's architecture, and also to whatever architecture incorporates the cpu. In a similar manner, the state from the pla's is passed from the pla monitor to the cpu architecture annotations, and also to the architecture annotations inside the pla architecture.

The design units required to support this visibility are shown in Figure 20.3. Assume an entity **pla** exists containing VAL annotations. Three design units are generated; two entity declarations and an architecture. The architecture (named **MONITOR**) contains the VHDL translation of the VAL annotations that appeared in the entity declaration. This includes the annotations which maintain the entity's state model. The ports of architecture **MONITOR** are the same as for entity **pla** with the addition of an **out** port of the same type as the entity's state model. This **out** port is used to provide visibility over the state of components of type **pla** to any annotations within any architecture that instantiates

[3]The value placed on a port by an entity does not necessarily equal the value on a signal connected to that port because bus resolution may come into play.

[4]In VHDL, a port of mode out is not readable within the architecture. Therefore assertions about out mode ports cannot be made in VHDL.

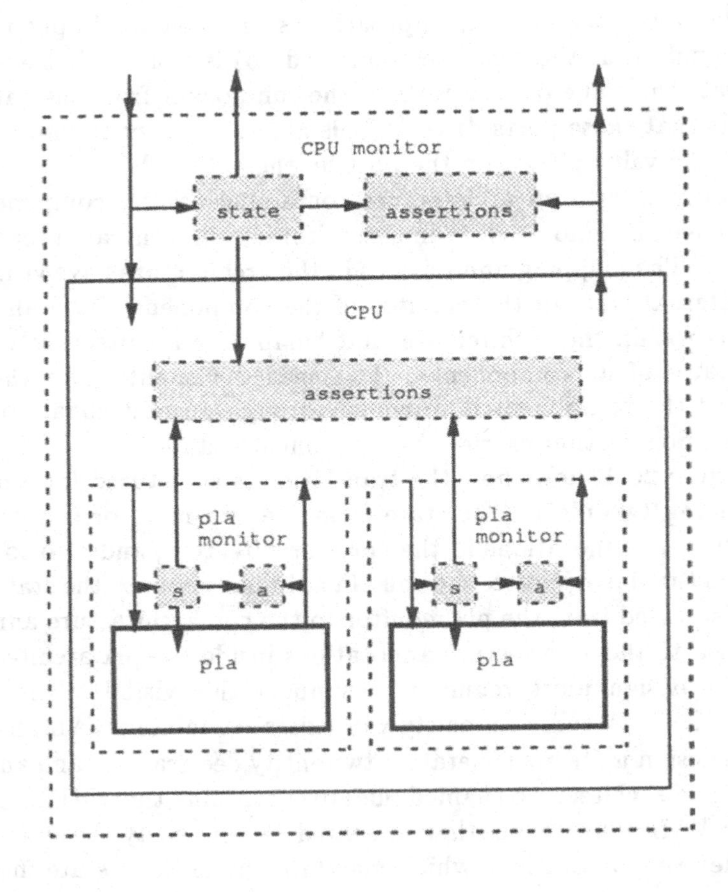

Figure 20.2: Schema of Architecture Annotation Translation

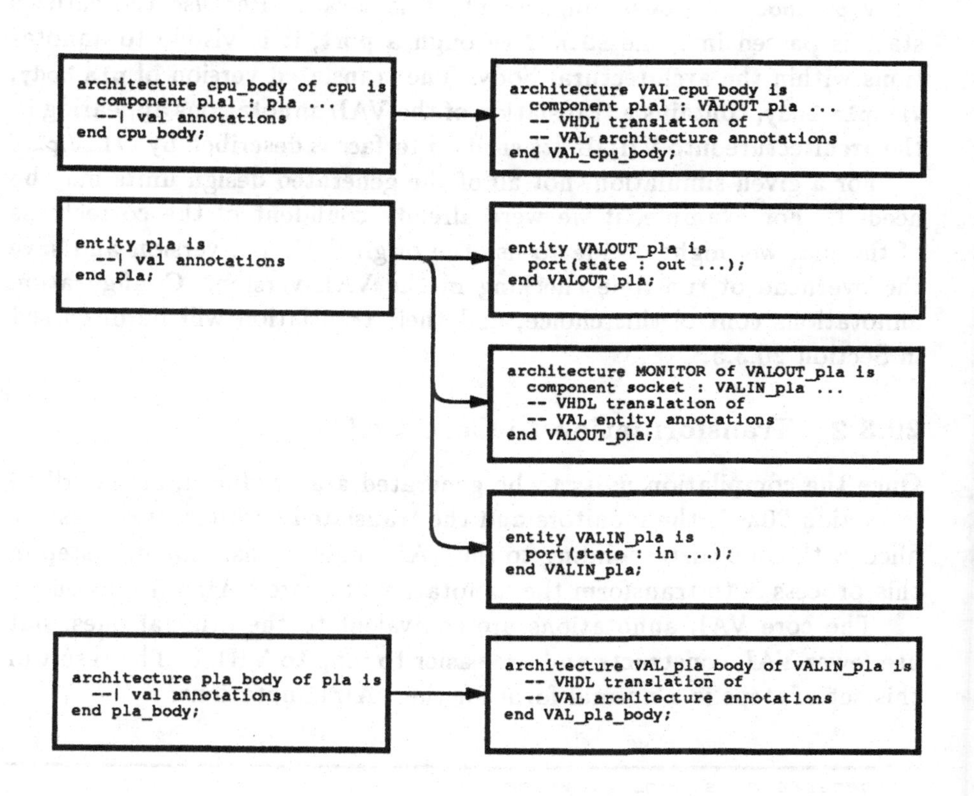

Figure 20.3: Relationship Between Design Units

a component of type **pla**. The generated entity **VALOUT_pla** declares the entity for **MONITOR**.

Architecture **MONITOR** contains a component **SOCKET** having the same ports as entity **pla** with the addition of an **in** port of the same type as the entity's state model. A translated version of the original architecture body **pla_body** of **pla** is plugged into this socket. Because the entity's state is passed into the **SOCKET** through a port, it is visible to annotations within the architectural body. The translated version of **pla_body**, **VAL_pla_body**, contains a translation of the VAL annotations appearing in the architecture into VHDL. Its entity interface is described by **VALIN_pla**.

For a given simulation, not all of the generated design units may be needed. For example, if we were already confident of the correctness of the pla, we might decide to use the original VHDL version and save the overhead of run-time checking in the VAL version. Configuration annotations control this choice, and their translation will be discussed in Section 20.3.5.

20.3.2 Transformation to Core VAL

Once the compilation units to be generated are produced, as described in section 20.3.1, the monitors and the translated architectures must be filled with code corresponding to the VAL annotations. The first step in this process is to transform the annotations to core VAL annotations.

The core VAL annotations are equivalent to the original ones, but use fewer VAL constructs and are easier to map to VHDL. The result of this set of steps is a normal form for the VAL annotations:

```
generation_scheme generate
  ...
  generation_scheme generate
    when expr then
      drive_process;
    end when;
  end generate;
  ...
end generate;
...
generation_scheme generate
  ...
```

```
    generation_scheme generate
      assertion_process;
    end generate;
    ...
  end generate;
```

Annotations consist of a list of **generate** statements, each of which executes in parallel. Each **generate** statement may contain exactly one nested statement. There may be a sequence of **generate** statements surrounding a single **when** statement or assertion process. The **when** statement guards a single drive process. The **generate** and/or **when** statement may be empty. That is, a **when** statement may appear at the top level, in parallel with the **generate** statements. This is equivalent to the same **when** statement nested within a **generate** statement that generates one copy of the statement it surrounds. Similarly, a missing **when** statement is equivalent to a **when** statement where the guard expression is the constant **TRUE**. Thus it will prove sufficient to translate VAL to VHDL if we can translate the above normal form. Although it is not shown in the previous picture, VAL expressions must also be converted to a normal form.

The transformation steps required to produce the normal form are:

1. Elimination of derived syntax. **Select**, **macro**, and complex forms of **when** statements are eliminated.

2. Separation of statements. All nested statement lists are eliminated. The only remaining statement list is at the top level.

3. Normalization of **generate** statements. **Generate** statements are moved to the outermost levels.

4. Flattening of nested guards. Nested **when** statements are eliminated.

5. Normalization of time qualified expressions. Time qualified expressions are shifted so that the upper time bound is zero.

6. Normalization of drive processes. drive processes are shifted so that the change in state occurs at relative time zero.

7. Normalization of assertion statements. Guards surrounding assertion statements are eliminated and assertions are shifted so that they only reference past time.

8. Normalization of timed expressions. Timed expressions are placed in a normal form.

This approach has several advantages over an approach where more complex VAL constructs are translated directly into VHDL:

1. These steps are independent of the target HDL, and may be performed regardless of the language to which VAL is translated.

2. The language-dependent part of the translation process is reduced in complexity, making it easier to translate VAL to other languages.

3. The semantics of more complex language constructs are implicitly defined in terms of other simpler VAL constructs

The following sections describe each of these steps in more detail.

20.3.2.1 Elimination of Derived Syntax

Derived syntax refers to language constructs that can be rewritten in terms of syntactically simpler language constructs. The rules that specify how to eliminate a language construct are known as *rewrite* rules since they specify how to rewrite one construct in terms of another.

The following VAL constructs are eliminated by this step:

- Select – The `select` statement activates one of a set of child statements based on the value of a selection expression. It can be re-written as a set of `when` statements:

```
select expr is
  in ce11 | ce12 | ... | ce1j => process_list1
  ...
  in cen1 | cen2 | ... | ceni => process_listn
  in others => process_list
end select;
```

```
==>

when ((ce11 = expr) or (ce12 = expr) or ... or (ce1j = expr)
  then process_list1
end when;
  ...
when ((cen1 = expr) or (cen2 = expr) or ... or (ceni = expr)
  then process_listn
end when;
when ((ce11 /= expr) and ... and (cen1 /= expr)
  then process_list
end when;
```

This, and the other rewrite rules, neglect semantic checking. If compile time semantic checks can guarantee no semantic errors, then the behavior of the rewritten expression is correct. Otherwise the transformation rule can be extended to include run-time semantic checking.

- Macro – A macro is a name for a list of parameterized statements. For every occurrence of the macro, the statements associated with the macro are copied, and the actual parameters substituted for the formal parameters. The rewrite rules must be performed recursively on the result of the expansion.

- Else and elsewhen – The syntactically more complex forms of the guarded statements are rewritten into simpler forms:

```
when expr1 then process_list1
  elsewhen expr2 then process_list2
  ...
  elsewhen exprn then process_listn
  else process_list
end when;

  ==>

when expr1 then process_list1 end when;
when ((not expr1) and expr2) then
  process_list1
end when;
```

```
...
when ((not expr1) and ... and (not exprn-1) and exprn) then
  process_listn
end when;
when ((not expr1) and ... and (not exprn)) then
  process_list
end when;
```

Once the rewrite rules have been performed recursively on the VAL description, only simple **when** statements (with no **else** parts), drive processes, and assertion processes remain.

20.3.2.2 Separation of Processes

At this point in the translation process, **select**, **macro**, and complex **when** statements have all been eliminated. We are now left with simple **when** statements (no **else** clauses), assertion processes, drive processes, and **generate** statements. Each **when** statement may, however, guard a list of statements. **When** statements may also still be nested. In addition, **generate** statements may generate a list of statements. In the next step, statement lists are broken apart so that each guarded statement guards exactly one statement and each **generate** statement generates exactly one statement. This simplifies the complexity of later steps in the translation algorithm.

Statement lists may occur within **when** statements and within **generate** statements. Statement lists within **when** statements are handled by the following rule:

```
when expr then
  process1;
    ...
  processn
end when;

  ==>

when expr then
  process1;
end when;
  ...
```

```
when expr then
  process_n;
end when;
```

Statement lists within **generate** statements are handled in a similar manner:

```
generation_scheme generate
  process_1;
    ...
  process_n
end generate;

  ==>

generation_scheme generate
  process_1;
end generate;
  ...
generation_scheme generate
  process_n;
end generate;
```

20.3.2.3 Normalization of Generate Statements

At this step in the translation, there are no remaining nested statement lists. But there may still be arbitrary nesting of **generate** statements and **when** statements. In this step, two functions are performed. First, all of the **generate** statements are moved to the outermost level of statement. This allows nested **when** statements appearing inside generate statements to be flattened into a single **when** statement by a later step.

```
when expr then
  generation_scheme generate
    ...
  end generate;
end when;

  ==>

generation_scheme generate
```

```
    when expr then
      ...
    end when;
  end generate;
```

Second, the names of the **generate** parameters are changed so that they are unique in the scope of the VAL annotations. The new unique name is substituted everywhere for the original name. Generate parameters are normally unique only within the block implicitly defined by the generate statement. When the statement is translated in a later stage to VHDL, the generate parameter names are used in such a way that they must be unique across all the annotations, not just the scope of the generate. Renaming the parameters now simplifies the translation process later.

20.3.2.4 Flattening of Nested Guards

At this step in the translation process, only **generate** statements, simple **when** statements, drive processes, and assertion processes remain. All process lists have been eliminated. A **when** statement now guards only a single statement, and a **when** statement never guards a **generate** statement. However, **when** statements may still be nested. In this step, nested **when** statements are elimitated:

```
when expr1 then
  when expr2 then
    process;
  end when;
end when;

  ==>

when (expr1 and expr2) then
  process;
end when;
```

After completing this step, the VAL description has the following structure:

```
generation_scheme generate
  ...
  generation_scheme generate
    when expr then
      drive_or_assertion_process;
    end when;
  end generate;
  ...
end generate;
...
generation_scheme generate
  ...
  generation_scheme generate
    when expr then
      drive_or_assertion_process;
    end when;
  end generate;
  ...
end generate;
```

The **generate** statement may be missing, which we may think of as a generate over a range of `1`, and the **when** statement may be missing, which we may think of as a **when** with a guard of `TRUE`. This as the structure of the normal form described earlier, but the expressions appearing in it must still be normalized.

20.3.2.5 Normalization of time qualified expressions

VHDL does not allow references to future time. That is, in VHDL one can only refer to a delayed version of a signal. Therefore all VAL expressions which refer to the future values of signals must be shifted so that the expression refers to only past values of signals. The first step in this process is normalizing the upper bound of time qualified expressions to 0.

Time qualified expressions may also contain upper bounds less than 0. Although expressions of this form may be expressed in VHDL without normalizing them, they are also normalized. As will be seen later, this simplifies the translation to VHDL since it may be assumed that the upper bound is always 0.

The normalization of time qualified expressions consists of two steps:

1. **Generalize defaults** – Default time references are added to every time qualified expression:

```
expr during T1

  ==>

expr during[-T1,0]
```

2. **Set upper bounds** – The upper bound of all time qualified expressions is set to zero:

```
expr during[T1,T2]

  ==>

expr[T2] during[T1-T2,0]
```

20.3.2.6 Normalization of Drive Statements

Since time in VAL is relative, the reference point of the entire description can be shifted in time such that all references are to the past. This facilitates the translation to VHDL since only the past value of a signal can be referenced in VHDL. All drive processes must be shifted in time so that the assignment to state occurs at time zero. This is accomplished by rewriting all guarded drive processes according to the following rule:

```
when expr1 then
  name[T] <- expr2;
end when;

  ==>

when expr1[-T] then
  name[0] <- expr2[-T]
end when;
```

If a drive process is unguarded, then we apply the following rule instead:

```
name[T] <- expr2;

  ==>

name[0] <- expr2[-T]
```

Alternatively, one can consider this as the case in which the drive process is guarded by a **when** statement in which the guard expression is the constant **TRUE**.

There are two things worth noting here: (1) the target of a drive process must be a component of the state, and, (2) normalization of timed expressions is unaffected by **generate** statement. The later is true because a **generate** statement cannot make a reference to time.

20.3.2.7 Normalization of Assertion Processes

Normalization of assertion processes involves two steps. First, any **when** statement guarding the assertion process is eliminated. (The previous steps have assured that there is at most one such guard.) Second, all timed expressions in the assertion are shifted so that they refer to past values of time. This step is performed for all assertion statements regardless of the flavor. The different flavors of assertions are handled later as part of code generation.

Eliminating a guard is a simple procedure:

```
when expr1 then
  assert expr2 report expr4 severity expr3;
end when;

  ==>

assert expr2 or not expr1 report expr4 severity expr3;
```

This is consistent with the definition of guarded assertion statements from Chapter 3. It holds for all flavors of assertions.

In a guarded drive process, the reference point is the point at which the assignment to the state takes place. Normalization for a drive process simply requires setting the point at which the assignment to state occurs to zero. Normalization of an assertion process requires that all references to time everywhere in the assertion process be to past values. This can be guaranteed in an assertion statements by computing the furthest forward reference needed to determine if an assertion holds, and shifting the entire assertion by that value. The amount of the shift is known as **TMAX**:

```
assert expr2 report expr4 severity expr3;

  ==>

signal TMAX : time;

TMAX <= max_time(assertion_process);

assert expr2[-TMAX]
     report expr4[-TMAX] severity expr3[-TMAX];
```

Where **TMAX** is an expression of type **TIME** computed according to the following algorithm:

```
max_time(when_statement)        = max(0,
                                      max_time(guard_expr),
                                      max_time(guarded_statement))
max_time(assertion_process)   = max(0,
                                      max_time(assert_expr),
                                      max_time(report_expr),
                                      max_time(severity_expr))
max_time(drive_process)       = max(0,
                                      max_time(target_expr),
                                      max_time(drive_expr))
max_time(name[T])             = T
max_time(expr[T])             = T + max_time(expr)
max_time(expr during[T1,0])    = max_time(expr)
max_time(expr1 op expr2)       = max(max_time(expr1),
                                       max_time(expr2))
max_time(op expr)              = max_time(expr)

max(expr1,...,exprn) = max(expr1,max(expr2,...,exprn))
max(expr1,expr2)     = expr1 if (expr1 >= expr2) else expr2
```

Intuitively, TMAX is the time that we have to wait to have information to evaluate a process.

Although not all of these rules may be used in normalizing assertion processes, they have been included in this definition because they will be important in the test for causality as part of the code generation scheme for a drive process described in Section 20.3.3.3.

20.3.2.8 Normalization of Timed Expressions

For the purposes of translating VAL to VHDL, it is important not only that the reference point of each guarded process be normalized to 0, but also that subexpressions not refer to future time. For example, the obvious translation for a timed expression is to replace it by a signal S:

```
expr[T]

  ==>

S <= expr after -T;
```

In order for this to succeed, the time T must be less than or equal to 0. That is, it must reference past values of the expression. The translation steps leading up to this point have all been applying negative shifts to expressions to ensure this, but the shifts have been applied to only the outer level of expressions, and not to any nested subexpressions. We must now apply the distributive rules for time shift to guarantee that time shifts in all subexpressions are negative. Four distributive rules are applied recursively:

```
(expr op expr)[T]          ==> (expr[T] op expr[T])
(expr[T1])[T2]             ==> expr[T1 + T2]
(op expr)[T]               ==> op(expr[T])
(expr during[T1,0])[T]     ==> (expr[T]) during[T1,0]
```

Repeated application of these rules will eventually result in a form for expressions in which shift operators are applied only to names and never to expressions.

If the expressions were part of an assertion process (or guarded assertion process), the time shift in all timed expressions are guaranteed to be less than 0 because a time shift of **-TMAX** was applied to the entire process during normalization of assertion statements. If the expression is part of a drive process, then we have not guaranteed that all time shifts are less than zero. In the case where not all the time shifts are less than zero, the expression is non-causal, and cannot be implemented in VHDL. Non-causal constructs are illegal in VAL and the code generation algorithm for drive statements detects this error. Assertion statements can never be non-causal since they never cause a change in state.

20.3.3 Code Generation

Once the preceding transformations have been applied to the VAL description, the code is in a canonical form characterized by:

- Only simple **when** statements with no nesting or else clauses.
- Time references rescaled relative to zero.
- Upper bound of time qualified expressions set to zero.
- One drive process or assertion process per when statement.
- Each **when** statement surrounded by a sequence of generate statements.

Each **generate** statement may contain exactly one nested statement. There are two cases: (1) a sequence of **generate** statements surrounds a single when statement which guards a single assertion process, or, (2) a sequence of **generate** statements surrounds an assertion process. The drive or assertion process will be referred to as the *base* statement. In addition to these two cases of statements, timed expressions and time qualified expressions must also be translated into the corresponding VHDL.

The transformation of the core VAL Annotations into VHDL are performed in four steps:

1. Translation of timed expressions.
2. Translation of time qualified expressions.
3. Translation of drive processes.

4. Translation of assertion processes.

The following sections describe the translation of each of these language constructs.

20.3.3.1 Translation of timed expressions

Recall that in an earlier translation step all time references were rescaled relative to the constant **TMAX**. Therefore all timed expressions must be less than zero; i.e. all timed expressions are delays. This can be modeled in VHDL by a signal assignment statement using *transport* delay.[5]

```
expr[T]

  ==>

signal S : expr_type;

S <= transport expr after -T;
```

All occurrences of the expression e[T] are then replaced with the signal S. All expressions are rewritten recursively until all timed expressions are eliminated. Transport delay is used to assure that no preemption [17] occurs on the signal. In VAL, once an assignment to a signal is made, it cannot be "undone."

The signal S is declared in the declaration part of the block enclosing the timed expression. Since the timed expression must appear in either a drive or assertion statement (or their guard), and the translation for drive and assertion statements begins a local block (see the following sections), S will be local to that block.

20.3.3.2 Translation of time qualified expressions

Recall that in an earlier step each time qualified expression was shifted in time such that its upper bound was zero. This can be translated

[5]The predefined VHDL attribute 'delayed() cannot be used for this because the argument of 'delayed() must be a globally static expression. (See Section 7.4 of [25].) Although the argument generated by the translation algorithm is a "run-time" constant, it is actually computed at elaboration time using functions defined in a VAL package. Therefore it does not meet the VHDL definition of globally static.

into VHDL as a check for stability over the most recent interval using a VHDL process.[6] An expression of the form

```
expr during[T,0]
```

is replaced by the signal S which is defined in VHDL in Figure 20.4. Note that T must be less than zero since the original during expression (before normalization) was required to have a lower bound less than its upper bound, and that the upper bound was subtracted from the lower bound to produce T.

```
signal S : boolean;

block
  signal gbel,gbel_delay : boolean;
  signal gbel_stable : boolean := true;
begin
  gbel <= expr;
  gbel_delay <= transport gbel after -T;
  process(gbel)
  begin
    if gbe_l'event then
      gbel_stable <= false;
      gbel_stable <= transport true after -T;
    end if;
  end process;
  S <= gbel_stable and gbel;
end block;
```

Figure 20.4: Translation of Time Qualified Expression

Whenever the expression changes value, the process is activated and sets a flag to false to indicate that the expression is not stable. The flag

[6] As with 'delayed(), the predefined VHDL attribute 'stable() cannot be used in the translation because the argument may not be a globally static expression as defined in VHDL. The argument may not be a globally static expression because the Transformer introduces function calls as part of the translation process. In effect, the transformer generates code to implement the 'stable() attribute itself.

is reset if the process is not activated (the expression does not change value) for `T` time units. Whenever the value of the expression changes, `S` is set to true if the expression is true and has been stable and true for the last T time units.

20.3.3.3 Translation of drive processes

In VAL/VHDL, the drive process can only be used to change the value of the entity state. After the previous transformations, there may be several guarded statements containing a drive process affecting the entity state or a component of the state. Only one of these, however, should be active at any point in time. Because VHDL requires that a signal may be the target of only a single concurrent signal assignment statement, all of the guarded statements that may influence the state are brought together into a single VHDL process. This process is sensitive to all of the signals that may influence the state, and checks that only a single assignment to state is active at any point in time.

Figure 20.5 shows the general form of the set of drive processes before applying the translation algorithm.

The result of applying the translation algorithm to Figure 20.5 is shown in Figure 20.6. A **generate** statement with a **for** generation scheme becomes a **for** statement, a **generate** statement with an **if** generation scheme becomes an **if** statement, a guarded statement becomes an **if** statement, and the drive itself becomes a signal assignment statement.

The translation is correct with these exceptions:

1. It does not check for multiple assignments to the same component of the state at the same time.

2. It does not check the causality of the assignments to state.

3. We have not described how to create the sensitivity list for the process.

In order to check (1) multiple assignment to the state, we must mark each component of the state as having been a target of an assignment when the state maintenance process is activated. If an attempt is made to assign to an already marked component of the state, then an error is generated. We use the following algorithm to check this:

```
for gen_parm11 generate
  if gen_cond1 generate
    ...
    for gen_parm1i generate
      when guard_expr1(gen_parm11,...,gen_parm1i) then
        state_comp1 <- drive_expr1(gen_parm11,...,gen_parm1i);
      end when;
    end generate;
    ...
end generate;
...
for gen_parmn1 generate
  if gen_condn generate
    ...
    for gen_parmnk generate
      when guard_exprn(gen_parmn1,...,gen_parmnk) then
        state_compn <- drive_exprn(gen_parmn1,...,gen_parmnk);
      end when;
    end generate;
    ...
end generate;
```

Figure 20.5: Normalized set of Drive Statements Before Translation

```
block
  process(...)
  begin
    for gen_parm11
      if gen_cond1
        ...
        for gen_parm1i
          if guard_expr1(gen_parm11,...,gen_parm1i) then
            state_comp1 <- drive_expr1(gen_parm11,...,gen_parm1i);
          end if;
        end for;
        ...
    end for;
    ...
    for gen_parmn1
      if gen_condn
        ...
        for gen_parmnk
          if guard_exprn(gen_parmn1,...,gen_parmnk) then
            state_compn <- drive_exprn(gen_parmn1,...,gen_parmnk);
          end if;
        end for;
        ...
    end for;
  end process;
end block
```

Figure 20.6: Translation of Drive Statements to VHDL

1. Declare a variable in the state maintenance process with a component of type **boolean** for every component of the type of the state. Call this variable **state_mutex**.

2. Set all components of **state_mutex** to **FALSE** at the start of the state maintenance process.

3. Replace every signal assignment to a component of state in the state maintenance process by an if statement that first checks if the component of **state_mutex** associated with the state component is **false** before doing the assignment. If the component of **state_mutex** is true, then multiple assignments to this component of state have occurred, and an error message is generated.

4. Set the component of **state_mutex** associated with the state component to **TRUE**. If the component is a selected or indexed component, set **state_mutex** for all parents of the component to true.

In order to check causality (the second item in our previous list), we use the function **max_time** specified in Section 20.3.2.7. For each guarded drive process, we assert that **max_time** of the guarding **when** statement is equal to zero. Recall that the drive process was normalized in a previous step so that the change in state occurs at relative time zero. If the drive process and its associated guard reference any time greater than zero, then the statement is non-causal.

For each guarded drive process, we have:

```
when expr then
  drive_process;
end when;

  ==>

assert max_time(when_statement) = 0 else
    report "Non-causal change in state";
```

Finally, we must generate the sensitivity list for the process. Intuitively, the process must be sensitive to any change in the generate parameters, guard, or assigned expression. Essentially this means that the process is sensitive to any name appearing in one of these expressions. However, we must be careful in generating the sensitivity list since the names may be indexed, selected, or attributed.

20.3.3.4 Translation of assertion processes

The translation of assertion processes is considerable simpler than the translation of drive processes. Translation of assertions occurs in two parts:

1. The **generate** statements surrounding an assertion are translated directly into VHDL.

2. The assertion itself is translated depending on the flavor of assertion.

There are four flavors of assertions in VAL: **assert**, **finally**, **sometime**, and **eventually**. Each of these assertions is translated into a VHDL process, the details of which depend on the particular flavor of assertion. Because the default severity level in VAL is **WARNING**, the translation must also set the severity level of generated VHDL assertions.

20.3.3.4.1 Assert The VAL **assert** process is translated directly into the VHDL **assert** statement:

```
assert expr1 report expr3 severity expr2;

  ==>

assert expr1 report expr3 severity expr2;
```

20.3.3.4.2 Finally The **finally** assertion is translated into a VHDL process that wakes up whenever a signal in the asserted expression changes. For the assertion:

```
finally test_expr
   report message_expr
   severity severity_expr;
```

```
val_finally : block
  signal next_time    : boolean;
  signal assert_expr : boolean;
begin
  assert_expr <= test_expr;
  process(assert_expr,next_time)
    variable first : boolean := true;
    variable oneb  : boolean := true;
  begin
    if (not assert_expr'event and
        not next_time'event) then
      next_time < not next_time after 1fs;
      first := false;
    end if;
    if next_time'event then
      assert oneb
        report message_expr severity severity_expr;
      first := true;
    end if;
    if assert_expr'event then
      if (assert_expr /= oneb) then
        oneb := assert_expr;
        if first then
          next_time <= not next_time after 1fs;
          first := false;
        end if;
      end if;
    end if;
  end process;
end block val_finally;
```

Figure 20.7: Translation of Finally assertion

the corresponding VHDL process is given in Figure 20.7.

When the process wakes up because of a change in the **test_expr**, it sets itself to wake up at the first delta of the next time and remembers the value of the **test_expr**. When it wakes at the next time point, the value of the asserted expression will be the value it held at the end of all of the deltas in the previous time point. The process then tests it to see if the assertion holds.

20.3.3.4.3 Sometime The translation for the **sometime** flavor of assertion closely resembles that for **finally**. Whenever a signal in the test expression changes, a process wakes up and checks if the test expression is true. The process then sets itself to wake up on the first delta of the next simulated time. When it wakes up at the next simulated time, the process checks that the expression was true in at least one delta in the previous simulation cycle. The translation for **sometime** is given in Figure 20.8.

20.3.3.4.4 Eventually The **eventually** flavor of assertion is similar to **finally**, except that once the test expression goes true it must remain true during all deltas in the remainder of the time point. the translation is thus very similar to that for **finally**, with the addition that the process must check that the test expression never makes the transition from false to true and back to false at the same time point.

The translation for **eventually** is given in Figure 20.9.

20.3.3.5 Attributes

This boolean attribute may have a parameter, such as in **'changed(val)** or may not have one, such as in **'changed**.

If **X'changed** is TRUE, it means that X changed its value in the last VHDL simulation cycle. If **X'changed** is FALSE, it means that **X** did not change its value since the last VHDL simulation cycle. Thus, **X'changed** is transformed as:

```
X'changed

  ==>
```

```
val_sometime : block
  signal next_time, assert_expr : boolean := false;
begin
  assert_expr <= test_expr;
  process(assert_expr,next_time'transaction)
    variable oneb  : boolean := false;
    variable first : boolean := true;
  begin
    if (not assert_expr'event and
        not next_time'transaction'event) then
      next_time <= not next_time after 1 fs;
      first := false;
    end if;
    if next_time'event then
      assert oneb
        report message_expr severity severity_expr;
      first := true;
      oneb :=  false;
    end if;
    if (assert_expr'event or not next_time'event) then
      oneb := oneb or assert_expr;
      if (first and not assert_expr) then
        next_time <= not next_time after 1fs;
        first := false;
      end if;
    end if;
 end process;
end block val_sometime;
```

Figure 20.8: Translation of Sometime assertion

```
val_eventually : block
  signal next_time, assert_expr : boolean := false;
begin
  assert_expr <= test_expr;
  process(assert_expr,next_time)
    variable glitch : boolean := false;
    variable oneb   : boolean := false;
    variable first  : boolean := true;
  begin
    if (not assert_expr'event and not next_time'event) then
      next_time <= not next_time after 1fs;
      first := false;
    end if;
    if next_time'event then
      assert glitch
        report message_expr severity severity_expr;
      first := true;
      glitch := false;
      oneb := false;
    end if;
    if (assert_expr'event) then
      glitch := glitch or (oneb and not assert_expr);
      oneb := assert_expr;
      if (first and not oneb) then
        next_time <= not next_time after 1fs;
        first := false;
      end if;
    end if;
  end process;
end block val_eventually;
```

Figure 20.9: Translation of Eventually assertion

```
not X'stable
```

If **X'changed(expr)** is true, it means that **X** has changed its value to **expr** in the last simulation cycle. If **X'changed(expr)** is false, it means that **X** has not changed to **expr** in the last VHDL simulation cycle. Thus, **X'changed(expr)** is transformed as:

```
X'changed(expr)

  ==>

((not X'stable) and (X = expr))
```

20.3.4 Architecture Annotations

A VAL-annotated VHDL architecture is translated in a VHDL architecture. The prefix **val** is added to the original name of the architecture in order to produce a unique identifier. The expanded architecture belongs to the **valin_** entity that is produced by transforming the original entity. The **valin_** entity has a port **state** of mode in that provides the expanded architecture with visibility over the state of the Monitor.

The component declarations within the expanded architecture may contain, in addition to their original ports, a port called **ValState** of a type indicated by the associated VAL annotation. If such a VAL annotation is missing, the component declarations maintain their original ports.

The instances of components that have the additional **ValState** port, have this port connected to a signal named *InstanceName*_**State**. Also, all references to *InstanceName*.**State** are transformed into references to the signal *InstanceName*_**State**.

In order to make the out ports visible to annotations, a signal is declared for each out port.

20.3.5 Configuration Annotations

The VAL user can select the entities to be monitored with the VAL constructs **valentity** and **valarchitecture**.

- The **valentity** construct specifies that an entity is to be monitored using interface annotations.

- The **valarchitecture** construct specifies that an architecture is to be monitored using body annotations.

Both annotations appear inside a VHDL component configuration immediately following the binding indication. Component configurations appear in either architecture body or configuration declaration. Currently the two constructs is supported only in configuration declarations.

For each entity `E`, with an architecture `A`, the VAL transformer will produce two interfaces and four architectures:

1. **entity** `Valout_E`: entity `E` with an extra `ValState` out port;

2. **architecture** `StateMonitor` **of** `valout_E`: monitor with a socket for the `Valin_E` entity, state maintenance and assertion checking;

3. **architecture** `DummyMonitor` **of** `Valout_E`: monitor with a socket for the `Valin_E` entity, but neither state maintenance nor assertion checking;

4. **entity** `Valin_E`: entity `E` with an extra `ValState` in port;

5. **architecture** `Val_A` **of** `Valin_E`: the architecture `A` of entity `E` with an extra `ValState` out port has been added to all components (if they themselves have a state model), and where the VAL annotations have been translated into VHDL statements.

6. **architecture** `A` **of** `Valin_E`: the architecture `A` of entity `E`, where all components have been added an extra `ValState` out port, but without translation for VAL annotations.

Assuming now a component configuration:

```
    for ... use entity E(A);
```

The VAL transformer will always add an extra level in the component hierarchy. Depending on the specified configuration annotations, the transformer will pick from the available architectures. If the

valentity construct appears in the component configuration, the transformer will select the **StateMonitor** architecture. If not, it will select the **DummyMonitor**. If the **valarchitecture** construct appears in the component configuration, the transformer will select the **Val_A architecture**. If not, it will select the **A** architecture.

```
for ... use entity E(A);
    for A
        ...
    end for;
end for;

for ... use entity Valout_E (DummyMonitor);
    for DummyMonitor
        for actual : socket use entity Valin_E (A);
            for A
                ...
            end for;
        end for;
    end for;
end for;
```

```
for ... use entity E(A);
    --| Valentity;
    for A
        ...
    end for;
end for;

for ... use entity Valout_E (StateMonitor);
    for StateMonitor
        for actual : socket use entity Valin_E (A);
            for A
                ...
            end for;
        end for;
    end for;
end for;
```

```
for ... use entity E(A);
```

```
        --| Valentity;
        --| Valarchitecture;
        for A
            ...
        end for;
    end for;

    for ... use entity Valout_E (StateMonitor);
        for StateMonitor
            for actual : socket use entity Valin_E (Val_A);
                for Val_A
                    ...
                end for;
            end for;
        end for;
    end for;
```

Figure 20.10 provides an example where the original VHDL architecture is to be used in conjunction with the Val annotated entity.

```
    for all: DFlipFlop use
      entity DFlipFlop (simple);
      --| ValEntity;
    end for;

-- is transformed into

    for all: DFlipFlop use
      entity ValOut_DFlipFlop (StateMonitor);
      for StateMonitor
        for actual: Socket use
          entity Valin_DFlipFlop (Simple);
        end for;
      end for;
    end for;
```

Figure 20.10: Transformation of Configuration Using "ValEntity"

Figure 20.11 provides an example where the Val annotated architecture is to be used in conjunction with the Val annotated entity.

```
    for all: DFlipFlop use
      entity DFlipFlop (Simple);
      --| ValEntity;
      --| ValArchitecture;
    end for;

-- is transformed into

    for all: DFlipFlop use
      entity ValOut_DFlipFlop (StateMonitor);
      for StateMonitor
        for actual: Socket use
          entity ValIn_DFlipFlop (Val_Simple);
        end for;
      end for;
    end for;
```

Figure 20.11: Transformation of Configuration Using "ValEntity" and "ValArchitecture"

20.4 Summary

The VAL entity annotation is transformed into a Monitor that watches a socket in which an architecture of the annotated entity can be plugged. Both the socket and the Monitor are hosted in a generated VHDL architecture belonging to a `VALIN_` entity.

Whenever such an entity is used, the configuration specifies which architecture to use. In turn the configuration of the generated architecture specifies some other architecture for the actual "chip" to be plugged in the socket.

This chapter presented the transformation to VHDL of various VAL specific constructs into VHDL in order to implement the monitoring activity.

Part V

Appendix

Appendix A

Syntax Summary

Syntax rules are described using the BNF notation. Each syntactic category is denoted by using lower case words, some containing embedded underlines, for example: adding_operation. A bold faced word is used to denote a reserved word in the language. The part enclosed by square brackets "[" and "]" means that it is optional, and thus it can be omitted. The part enclosed by braces "{" and "}" may be iterated zero, once, or multiple times. A vertical bar "|" separates alternative items. When these symbols appear as part of the VAL syntax, they are placed within single quotes.

If the name of syntactic category starts with an italicized part, it is equivalent to the category name without the italicized part. The italicized part is intended to convey some semantics information. For example, *boolean_*expression and *numeric_*expression are both equivalent to expression alone.

When the whole name of a syntactic category is italicized with a preceding annotation indicator "--|", it means that all the text in that category consists of annotations and thus each text line should start with an annotation indicator.

A VAL syntax rule is often obtained from a VHDL syntax rule either by additional clauses or by extension of the clauses in the VHDL rule. In these cases a reference to the VHDL rule enclosed by three dots is placed on the left and right sides of the VHDL rule.

A.1 Lexical Elements

```
[11.1]
graphic_character ::= ...(same as VHDL)

[11.4]
abstract_literal ::= (same as VHDL)
character_literal ::= (same as VHDL)
string_literal ::= (same as VHDL)
bit_string_literal ::= (same as VHDL)
```

A.2 Syntax

```
[16.6]
actual_part ::= expression

[12.2]
Annotated_architecture_body ::=
    architecture identifier of entity_name is
        VAL_VHDL_architecture_declarative_part
    begin
        VAL_VHDL_architecture_statement_part
    end [ architecture_simple_name ] ;

[12.2]
Annotated_block_statement ::=
    block_label :
        block [ (guard_expression) ]
            block_header
            VAL_VHDL_block_declarative_part
        begin
            VAL_VHDL_block_statement_part
        end block [ block_label ] ;

[18]
Annotated_component_configuration ::=
    for component_specification
      [ use binding_indication ; ]
```

```
        [ --| state_model_map_annotation ; ]
        [ --| VAL_binding_specification ; ]
        [ block_configuration ]
    end for ;
```

[18]

```
Annotated_configuration_declaration ::=
    configuration identifier of entity_name is
       --| VAL_entity_binding_specification ;
        configuration_declarative_part
        block_configuration
    end [ configuration_simple_name ] ;
```

[18]

```
Annotated_configuration_item ::=
    block_configuration
  | component_configuration
  | Annotated_component_configuration
```

[12.1]

```
Annotated_VHDL_entity_declaration ::=
    entity identifier is
          VAL_VHDL_entity_header
          VAL_VHDL_entity_declarative_part
  [ begin
        VAL_VHDL_entity_statement_part ]
    end [ entity_simple_name ] ;
```

[12.2]

```
architecture_declarative_item ::= ...(same as VHDL)
```

[12.2]

```
architecture_statement ::= ...(same as VHDL)
```

[16.1]

```
assertion_flavor ::=
    assert
  | finally
  | sometime
  | eventually
```

[16.1]

```
assertion_statement ::=
    assertion_flavor boolean_extended_expression
  [ else ]
```

```
        [ report expression ]
        [ severity expression ] ;
```

[16.6]
association_element ::= [formal_part =>] actual_part

[16.6]
```
association_list ::=
    association_element { , association_element }
```

[14.2]
```
assume_declaration ::=
    assume boolean_expression
  [ else ]
      [ report expression ]
      [ severity expression ] ;
```

[18]
```
block_configuration ::=
    for block_specification
      { use_clause }
      { Annotated_configuration_item }
    end for ;
```

[12.2]
block_declarative_item ::= ...*(same as VHDL)*

[12.2]
block_header ::= ...*(same as VHDL)*

[12.2]
block_statement ::= ...*(same as VHDL)*

[16.4]
```
choice ::=
    simple_expression
  | discrete_range
  | component_simple_name
  | others
```

[18]
component_specification ::= ...*(same as VHDL)*

[18]
binding_indication ::= ...*(same as VHDL)*

```
[18]
component_configuration ::= ...(same as VHDL)

[12.2]
component_declaration ::=
    component identifier
      [ generic ( local_generic_list ) ]
      [ port ( local_port_list ) ]
      [ --| VAL_state_model_declaration ]
    end component ;

[18]
conf_state_model_map_annotation ::=
    for component_specification
                state_model_map_annotation

[18]
conf_VAL_binding_specification ::=
    for component_specification
                VAL_binding_specification


[18]
configuration_declarative_part ::= ...(same as VHDL)

[16.2]
drive_statement ::=
    extended_expression -> object_name [ time_indicator ] ;
  | object_name [ time_indicator ] <- extended_expression ;

[12.1]
entity_declarative_item ::= ...(same as VHDL)

[12.1]
entity_header ::= ...(same as VHDL)

[12.1]
entity_statement ::= ...(same as VHDL)

[15]
extended_expression ::=
      ...VHDL_expression ...
  | timed_expression
  | time_qualified_boolean_expression
```

```
[16.6]
formal_part ::= identifier

[16.5]
generate_statement ::=
    VAL_generation_scheme generate
        VAL_statement_list
    end generate ;

guarded_statement ::=
    when boolean_extended_expression then
        VAL_statement_list
  { elsewhen boolean_extended_expression then
        VAL_statement_list }
  [ else
        VAL_statement_list ]
    end when ;

[14.4]
identifier_list := ...(same as VHDL)

[14.4]
macro_body ::= VAL_statement_list

[16.6]
macro_call_statement ::=
    macro_name ( association_list ) ;

[14.4]
macro_declaration ::=
    macro_specification is
        macro_body
    end simple_name;

[14.4]
macro_parameter_list ::= identifier_list

[14.4]
macro_specification ::=
    macro simple_name [ ( macro_parameter_list ) ]

[16.7]
null_statement ::= null ;

[19.1]
```

```
package_body_declarative_item ::= ...(same as VHDL)

[19.1]
package_declarative_item ::= ...(same as VHDL)

[16.5]
parameter_specification ::= ... (Same as VHDL)

[15]
primary ::=
      ...VHDL_primary ...
   | ( extended_expression )

[16.4]
select_statement ::=
    select extended_expression is
        select_statement_alternative
      { select_statement_alternative }
    end select ;

[16.4]
select_statement_alternative ::=
    in choice
  { '|' choice } => VAL_statement_list

[14.4]
simple_name ::= ...(same as VHDL)

[15.2]
simple_qualified_boolean_expression ::=
    boolean_expression during time_interval
    boolean_expression within time_interval

[13.1]
state_model_declaration ::=
    state model [ identifier ] is
        subtype_indication := expression ;

[18]
state_model_map_annotation ::=
    state model map (type_conversion_function_name)

[13.1]
subtype_indication ::= ...(same as VHDL)

[16.2]
```

```
time_indicator ::= '[' time_expression ']'

[16.3]
[15.2]
time_interval ::=
      '[' time_expression, time_expression ']'
  |       time_expression

[15.2]
time_qualified_boolean_expression ::=
      simple_qualified_boolean_expression
  { logical_operator
      simple_qualified_boolean_expression }

[15.1]
timed_expression ::=
       ... VHDL_expression ...
  | expression '[' time_expression ']'



[14]
VAL_architecture_declarative_item ::=
      alias_declaration
  | macro_declaration
  | use_clause
  | conf_state_model_map_annotation
  | conf_VAL_binding_specification

[14]
VAL_architecture_statement ::=
      assert_statement
  | guarded_statement
  | select_statement
  | replicate_statement
  | macro_call_statement
  | null_statement

[18]
VAL_binding_specification ::=
      VAL_entity_binding_specification
  | valconfiguration

[18]
```

```
VAL_entity_binding_specification ::=
  [ valentity ] [ valarchitecture ]

[14]
VAL_entity_declarative_item ::=
    type_declaration
  | subtype_declaration
  | constant_declaration
  | alias_declaration
  | use_clause
  | state_model_declaration
  | macro_declaration
  | assume_declaration

[16.5]
VAL_generation_scheme ::=
    for generate_parameter_specification

[19.1]
VAL_package_body_declaration ::=
    package body package_simple_name is
        VAL_VHDL_package_body_declarative_part
    end [ package_simple_name ] ;

[19.1]
VAL_package_body_declarative_item ::=
    subprogram_declaration
  | subprogram_body
  | type_declaration
  | subtype_declaration
  | constant_declaration
  | alias_declaration
  | use_clause
  | macro_declaration

[19.1]
VAL_package_declaration ::=
    package identifier is
        VAL_VHDL_package_declarative_part
    end [ package_simple_name ] ;

[19.1]
VAL_package_declarative_item ::=
    subprogram_declaration
```

```
  | type_declaration
  | subtype_declaration
  | constant_declaration
  | alias_declaration
  | use_clause
  | macro_specification
```

[16]

```
VAL_statement ::=
        assertion_statement
  | guarded_statement
  | select_statement
  | drive_statement
  | replicate_statement
  | macro_call_statement
  | null_statement
```

[14.4]

```
VAL_statement_list ::= { VAL_statement }
```

[12.2]

```
VAL_VHDL_architecture_declarative_item ::=
    --| VAL_architecture_declarative_item
  | architecture_declarative_item
```

[12.2]

```
VAL_VHDL_architecture_declarative_part ::=
   { VAL_VHDL_architecture_declarative_item }
```

[12.2]

```
VAL_VHDL_architecture_statement ::=
    --| VAL_architecture_statement
  | architecture_statement
```

[12.2]

```
VAL_VHDL_architecture_statement_part ::=
  { VAL_VHDL_architecture_statement }
```

[12.2]

```
VAL_VHDL_block_declarative_item ::=
    --| VAL_architecture_declarative_item
  | block_declarative_item
```

[12.2]

```
VAL_VHDL_block_declarative_part ::=
```

```
  { VAL_VHDL_block_declarative_item }

[12.1]
VAL_VHDL_entity_declarative_item ::=
  --|VAL_entity_declarative_item
  | entity_declarative_item

[12.2]
VAL_VHDL_block_statement ::=
    --| VAL_architecture_statement
  | block_statement

[12.2]
VAL_VHDL_block_statement_part ::=
  { VAL_VHDL_block_statement }

[12.1]
VAL_VHDL_entity_declarative_part ::=
  { VAL_VHDL_entity_declarative_item }

[12.1]
VAL_VHDL_entity_header ::=
    entity_header
  [ VAL_state_model_declaration ]

[12.1]
VAL_VHDL_entity_statement ::=
    --| VAL_entity_statement
  | entity_statement

[12.1]
VAL_VHDL_entity_statement_part ::=
  { VAL_VHDL_entity_statement }

[19.1]
VAL_VHDL_package_body_declarative_part ::=
   VAL_VHDL_package_body_declarative_item

[19.1]
VAL_VHDL_package_body_declarative_item ::=
    --| VAL_package_body_declarative_item
  | package_body_declarative_item

[19.1]
VAL_VHDL_package_declarative_item ::=
```

```
    --| VAL_package_declarative_item
  | package_declarative_item
```

[19.1]

```
VAL_VHDL_package_declarative_part ::=
  { VAL_VHDL_package_declarative_item }
```

Appendix B

CPU : VHDL description

B.1 One bit alu

```
use work.cpu_p.all ;

entity alu1 is

    port ( pwr, gnd : in bit := '0' ;
           a_i, b_i : in bit := '0' ;
           op_i : in bit_vector (1 to 2) := B"00" ;
           c_o : out bit := '0' ) ;

begin

--| when not (  a_i'event
--|          or b_i'event
--|          or c_o'event
--|          ) during [-20ns,0] then
--|     select op_i is
--|     in B"00" => assert c_o = a_i and b_i
--|                 report "alu1 : c_o /= a_i and b_i" ;
--|     in B"01" => assert c_o = a_i or b_i
--|                 report "alu1 : c_o /= a_i or b_i" ;
--|     in B"10" => assert c_o = not b_i
--|                 report "alu1 : c_o /= not b_i" ;
--|     in B"11" => assert c_o = a_i xor b_i
```

```
--|                    report "alu1 : c_o /= a_i xor b_i";
--|        end select ;
--| end when ;

end alu1 ;

architecture structure of alu1 is

    signal nota_s, notb_s : bit := '1' ;
    signal notop_s, notop2_s : bit_vector (1 to 2) := B"11" ;
    signal nand21_s, nand22_s,
           nand23_s, nand24_s : bit := '1' ;

begin

    nota_s <= not a_i after 2ns ;
    notb_s <= not b_i after 2ns ;
    notop_s <= not op_i after 2ns ;
    nand21_s <= not (a_i and b_i and notop_s(1)) after 3ns ;
    nand22_s <= not (nota_s and b_i and op_i(2)) after 3ns ;
    nand23_s <= not (a_i and notb_s
                        and notop_s(2)) after 3ns ;
    nand24_s <= not (notb_s and op_i(1)
                           and notop_s(2)) after 3ns ;
    c_o <= not (nand21_s and nand22_s and
                nand23_s and nand24_s) after 3ns ;

end structure ;
```

B.2 16 bit alu

```
use work.cpu_p.all ;

entity alu16  is

    port ( pwr, gnd : in bit := '0' ;
           a_i, b_i : in bit_vector (1 to 16) := X"0000" ;
           op_i : in bit_vector (1 to 2) := B"00" ;
           c_o : out bit_vector (1 to 16) := X"0000" ) ;
```

```
begin

--| when not (a_i'event or b_i'event
--|                       or c_o'event) during [-20ns,0] then
--|     select op_i is
--|     in B"00" => assert c_o = a_i and b_i
--|                 report "alu16  : c_o /= a_i and b_i" ;
--|     in B"01" => assert c_o = a_i or b_i
--|                 report "alu16  : c_o /= a_i or b_i" ;
--|     in B"10" => assert c_o = not b_i
--|                 report "alu16  : c_o /= not b_i" ;
--|     in B"11" => assert c_o = a_i xor b_i
--|                 report "alu16  : c_o /= a_i xor b_i" ;
--|     end select ;
--| end when ;

end alu16  ;



architecture structure of alu16  is

    component alu1_c
        port ( pwr, gnd : in bit := '0' ;
               a_i, b_i : in bit := '0' ;
               op_i : in bit_vector (1 to 2) := B"00" ;
               c_o : out bit := '0' ) ;
    end component ;

begin

    l : for i in 1 to 16 generate
        alu : alu1_c
            port map (pwr => pwr,
                      gnd => gnd,
                      a_i => a_i(i),
                      b_i => b_i(i),
                      op_i => op_i,
                      c_o => c_o(i)) ;
    end generate ;

end structure ;
```

B.3 One bit buffer

```
use work.cpu_p.all ;

entity buf1 is

    port ( pwr, gnd : in bit := '0' ;
           d_i : in bit := '0' ;
           OE_i : in bit := '0' ;
           d_o : out bit := '0' ) ;

begin

--| when not (d_i'event and
--|           OE_i'event) during [-10ns, 0ns] then
--|    when (OE_i = '1') then
--|        assert d_i = d_o report "buf : d_i /= d_o" ;
--|    else
--|        assert d_o = '0' report "buf : d_o /= '0'" ;
--|    end when ;
--| end when ;

end buf1 ;


architecture structure of buf1 is

    signal notd_s : bit := '1' ;

begin

    notd_s <= not (d_i and OE_i) after 3ns ;
    d_o <= not notd_s after 2ns ;

end structure ;
```

B.4 12 bit buffer

```
use work.cpu_p.all ;

entity buf12 is

    port ( pwr, gnd : in bit := '0' ;
           d_i : in bit_vector (1 to 12) := X"000" ;
           OE_i : in bit := '0' ;
           d_o : out bit_vector (1 to 12) := X"000" ) ;

begin

--| when not (d_i'event and
--|             OE_i'event) during [-10ns, 0ns] then
--|    when (OE_i = '1') then
--|        assert d_i = d_o report "buf : d_i /= d_o" ;
--|    else
--|        assert d_o = X"000" report "buf : d_o /= X000" ;
--|    end when ;
--| end when ;

end buf12 ;


architecture structure of buf12 is

    component buf1_c
        port ( pwr, gnd : in bit := '0' ;
               d_i : in bit := '0' ;
               OE_i : in bit := '0' ;
               d_o : out bit := '0' ) ;
    end component ;

begin

    l : for i in 1 to 12 generate
        b : buf1_c
            port map ( pwr => pwr,
                       gnd => gnd,
                       d_i => d_i(i),
                       OE_i => OE_i,
                       d_o => d_o(i) ) ;
    end generate ;

end structure ;
```

B.5 16 bit buffer

```
use work.cpu_p.all ;

entity buf16 is

    port ( pwr, gnd : bit := '0' ;
           d_i : in bit_vector (1 to 16) := X"0000" ;
           OE_i : in bit := '0' ;
           d_o : out bit_vector (1 to 16) := X"0000" ) ;

begin

--| when not (d_i'event and
--|           OE_i'event) during [-10ns, 0ns] then
--|    when (OE_i = '1') then
--|        assert d_i = d_o report "buf : d_i /= d_o" ;
--|    else
--|        assert d_o = X"0000" report "buf : d_o /= X0000" ;
--|    end when ;
--| end when ;

end buf16 ;


architecture structure of buf16 is

    component buf1_c
        port ( pwr, gnd : in bit := '0' ;
               d_i : in bit := '0' ;
               OE_i : in bit := '0' ;
               d_o : out bit := '0' ) ;
    end component ;

begin

    l : for i in 1 to 16 generate
        b : buf1_c
            port map ( pwr => pwr,
```

```
                              gnd => gnd,
                              d_i => d_i(i),
                              OE_i => OE_i,
                              d_o => d_o(i) ) ;
        end generate ;

    end structure ;
```

B.6 CPU

```
use work.cpu_p.all ;

entity cpu is

    port ( pwr, gnd : in bit := '0' ;
           clk_i, rst_i : in bit := '0' ;
           run_i, clr_i : in bit := '0' ;
           d_i : in bit_vector (1 to 16) := X"0000" ;
           ir_i : in bit_vector (1 to 16) := X"0000" ;

           d_o : out bit_vector (1 to 16) := X"000" ;
           addr_o : out bit_vector (1 to 12) := X"0000" ;
           read_o, write_o : out bit := '0'
         ) ;

    -- Annotation

--| STATE model is state_model := ( if1,
--|                                  (X"0000", X"0000", X"0000", X"0000"),
--|                                   X"0000", X"0000"
--|                                 ) ;


--| macro instr_fetch_1_state is
--|     when state.st = if1 then
--|         when clr_i = '1' then
--|             state.st <- if2 ;
--|         elsewhen run_i = '1' then
--|             state.st <- if3 ;
```

```
--|          end when ;
--|      end when ;
--| end instr_fetch_1_state ;

--| macro instr_fetch_2_state is
--|     when state.st = if2 then
--|         state.ir <- "0000";
--|         state.st <- if4 ;
--|     end when ;
--| end instr_fetch_2_state ;

--| macro instr_fetch_3_state is
--|     when state.st = if3 then
--|         state.ir <- ir_i ;
--|         state.st <- if4 ;
--|     end when ;
--| end instr_fetch_3_state ;

--| macro instr_fetch_4_state is
--|     when state.st = if4 then
--|         when run_i = '1' then
--|             select op is
--|                 in X"0" => state.st <- if1 ;
--|                 in X"1" => state.st <- ex1 ;
--|                 in X"2" => state.st <- ld ;
--|                 in X"3" => state.st <- st ;
--|             end select ;
--|         end when ;
--|     end when ;
--| end instr_fetch_4_state ;

--| macro execute_1_state is
--|     when state.st = ex1 then
--|         select ex is
--|             in X"0" => state.acc <- state.reg (intg (r1)) and
--|                                     state.reg (intg (r2)) ;
--|             in X"1" => state.acc <- state.reg (intg (r1)) or
--|                                     state.reg (intg (r2)) ;
--|             in X"2" => state.acc <- not state.reg (intg (r2))
--|             in X"3" => state.acc <- state.reg (intg (r1)) xor
--|                                     state.reg (intg (r2)) ;
--|         end select ;
--|         state.st <- ex2 ;
--|     end when ;
--| end execute_1_state ;

--| macro execute_2_state is
```

```
--|       when state.st = ex1 then
--|           state.reg (intg (r2)) <- state.acc ;
--|           state.st <- if1 ;
--|       end when ;
--|   end execute_2_state ;

--|   macro load_state is
--|       when state.st = ld then
--|           select r1 is
--|               in X"0" => state.reg0 <- d_i ;
--|               in X"1" => state.reg1 <- d_i ;
--|               in X"2" => state.reg2 <- d_i ;
--|               in X"3" => state.reg3 <- d_i ;
--|           end select ;
--|           state.st <- if1 ;
--|       end when ;
--|   end load_state ;

--|   macro store_state is
--|       when state.st = st then
--|           state.st <- if1 ;
--|       end when ;
--|   end store_state ;


      -- annotation state update
begin


--| when rst_i then
--|     state.st <- if1 ;
--| else
--|     when clk_i'changed('1') then
--|         instr_fetch_1_state ;
--|         instr_fetch_2_state ;
--|         instr_fetch_3_state ;
--|         instr_fetch_4_state ;
--|         execute_1_state ;
--|         execute_2_state ;
--|         load_state ;
--|         store_state ;
--|     end when ;
--| end when ;
```

```
        -- assertions

    --| when clk_i'changed ('1') then

    --|     when state.st = ld then
    --|         assert read_o = '1' ;
    --|         assert addr_o = state.ir (5, 16) ;
    --|     end when ;

    --|     when state.st = st then
    --|         assert write_o = '1' ;
    --|         assert addr_o = state.ir (5, 16) ;
    --|         assert d_o = state.reg (intg (r1)) ;
    --|     end when ;

    --| end when ;

    end cpu ;


    architecture structure of cpu is

        signal alu_s : bit_vector (1 to 16) := X"0000" ;
        signal ir_s : bit_vector (1 to 16) := X"0000" ;
        signal d_s : bit_vector (1 to 16) := X"0000" ;
        signal d_s1 : bit_vector (1 to 16) := X"0000" ;
        signal d_s2 : bit_vector (1 to 16) := X"0000" ;
        signal irCE_s,
               accCE_s,
               reg0CE_s,
               reg1CE_s,
               reg2CE_s,
               reg3CE_s,
               accOE_s,
               dinOE_s,
               irRST_s,
               addrOE_s,
               writeOE_s : bit := '0' ;

        signal notwrite_s : bit := '1' ;

        signal mux1_s1, mux2_s1 : bit_vector (1 to 16) := X"0000" ;
        signal mux1_s2, mux2_s2 : bit_vector (1 to 16) := X"0000" ;
        signal mux1_s3, mux2_s3 : bit_vector (1 to 16) := X"0000" ;
```

```
                                       X"0000", X"0000") ;
signal mux1_t, mux2_t : bit_vector (1 to 16) := X"0000" ;
signal r1_s, r2_s : bit_vector (1 to 2) := B"00" ;

signal regOE1_s, regOE2_s : bit_vector (0 to 3) := X"0" ;

component pla_c
   port ( pwr, gnd : in bit := '0' ;
          op_i : in bit_vector (1 to 2) := B"00" ;
          r1_i : in bit_vector (1 to 2) := B"00" ;
          r2_i : in bit_vector (1 to 2) := B"00" ;
          clr_i : in bit := '0' ;
          run_i : in bit := '0';
          rst_i : in bit := '0';
          clk_i : in bit := '0';
          irCE_o : out bit := '0';
          accCE_o : out bit := '0';
          reg0CE_o : out bit := '0';
          reg1CE_o : out bit := '0';
          reg2CE_o : out bit := '0';
          reg3CE_o : out bit := '0';

          reg0OE1_o : out bit := '0';
          reg1OE1_o : out bit := '0';
          reg2OE1_o : out bit := '0';
          reg3OE1_o : out bit := '0';

          reg0OE2_o : out bit := '0';
          reg1OE2_o : out bit := '0';
          reg2OE2_o : out bit := '0';
          reg3OE2_o : out bit := '0';

          accOE_o : out bit := '0';
          dinOE_o : out bit := '0';
          irRST_o : out bit := '0';
```

```
                addrOE_o : out bit := '0';
                readOE_o : out bit := '0';
                writeOE_o : out bit := '0' ) ;
        --| state model is states := ifl ;
end component ;

component ref16_c
        port ( pwr, gnd : in bit := '0' ;
                d_i : in bit_vector (1 to 16) := X"0000" ;
                rst_i : in bit := '0' ;
                clk_i : in bit := '0' ;
                CE_i : in bit := '0' ;
                OE_i : in bit := '0' ;
                d_o : out bit_vector (1 to 16) := X"0000" ) ;
        --| state model is integer range 0 to 65535 := 0 ;
end component ;

component reg16_c
        port ( pwr, gnd : in bit := '0' ;
                d_i : in bit_vector (1 to 16) := X"0000" ;
                rst_i : in bit := '0' ;
                clk_i : in bit := '0' ;
                CE_i : in bit := '0' ;
                OE1_i : in bit := '0' ;
                d1_o : out bit_vector (1 to 16) := X"0000" ;
                OE2_i : in bit := '0' ;
                d2_o : out bit_vector (1 to 16) := X"0000" ) ;
        --| state model is integer range 0 to 65535 := 0 ;
end component ;

component alu16_c
        port ( pwr, gnd : in bit := '0' ;
                a_i, b_i : in bit_vector (1 to 16) := X"0000" ;
                op_i : in bit_vector (1 to 2) := B"00" ;
                c_o : out bit_vector (1 to 16) := X"0000" ) ;
end component ;

component buf16_c
        port ( pwr, gnd : in bit := '0' ;
                d_i : in bit_vector (1 to 16) := X"0000" ;
                OE_i : in bit := '0' ;
                d_o : out bit_vector (1 to 16) := X"0000" ) ;
end component ;
```

```
component buf12_c
    port ( pwr, gnd : in bit := '0' ;
           d_i : in bit_vector (1 to 12) := X"000" ;
           OE_i : in bit := '0' ;
           d_o : out bit_vector (1 to 12) := X"000" ) ;
end component ;

component buf1_c
    port ( pwr, gnd : in bit := '0' ;
           d_i : in bit := '0' ;
           OE_i : in bit := '0' ;
           d_o : out bit := '0' ) ;
end component ;

component or16_c
    port ( pwr, gnd : in bit := '0' ;
           d1_i : in bit_vector (1 to 16) := B"00" ;
           d2_i : in bit_vector (1 to 16) := B"00" ;
           d_o : out bit_vector (1 to 16) := B"00" ) ;
end component ;

component orr16_c
    port ( pwr, gnd : in bit := '0' ;
           d1_i : in bit_vector (1 to 16) := B"00" ;
           d2_i : in bit_vector (1 to 16) := B"00" ;
           d3_i : in bit_vector (1 to 16) := B"00" ;
           d4_i : in bit_vector (1 to 16) := B"00" ;
           d_o : out bit_vector (1 to 16) := B"00" ) ;
end component ;


begin

    DIB : buf16_c
        port map (pwr => pwr,
                  gnd => gnd,
                  d_i => d_i,
                  OE_i => dinOE_s,
                  d_o => d_s1 ) ;


    l : for i in 0 to 3 generate
        R : reg16_c
            port map (pwr => pwr,
                      gnd => gnd,
```

```
                         d_i => d_s,
                         rst_i => gnd,
                         clk_i => clk_i,
                         CE_i => reg0CE_s,
                         OE1_i => regOE1_s (i),
                         d1_o => mux1_s,
                         OE2_i => regOE2_s (i),
                         d2_o => mux2_s) ;
end generate ;

O1 : orr16_c
    port map (pwr => pwr,
              gnd => gnd,
              d1_i => mux1_s1,
              d2_i => mux1_s2,
              d3_i => mux1_s3,
              d4_i => mux1_s4,
              d_o => mux1_s) ;

O2 : orr16_c
    port map (pwr => pwr,
              gnd => gnd,
              d1_i => mux2_s1,
              d2_i => mux2_s2,
              d3_i => mux2_s3,
              d4_i => mux2_s4,
              d_o => mux2_s) ;

ACC : ref16_c
    port map (d_i => alu_s,
              rst_i => gnd,
              clk_i => clk_i,
              CE_i => accCE_s,
              OE_i => accOE_s,
              d_o => d_s2) ;

O0 : or16_c
    port map (pwr => pwr,
              gnd => gnd,
              d1_i => d_s1,
              d2_i => d_s2,
              d_o => d_s) ;
```

```
A : alu16_c
    port map (pwr => pwr,
              gnd => gnd,
              a_i => mux1_s,
              b_i => mux2_s,
              op_i => ir_s (7 to 8),
              c_o => alu_s) ;

DOB : buf16_c
    port map (pwr => pwr,
              gnd => gnd,
              d_i => mux1_s,
              OE_i => writeOE_s,
              d_o => d_o) ;

C : pla_c
    port map (pwr => pwr,
              gnd => gnd,
              op_i => ir_s (1 to 2),
              r1_i => ir_s (3 to 4),
              r2_i => ir_s (5 to 6),
              clr_i => clr_i,
              run_i => run_i,
              rst_i => rst_i,
              clk_i => clk_i,
              irCE_o => irCE_s,
              accCE_o => accCE_s,

              reg0CE_o => reg0CE_s,
              reg1CE_o => reg1CE_s,
              reg2CE_o => reg2CE_s,
              reg3CE_o => reg3CE_s,

              reg0OE1_o => regOE1_s(1),
              reg1OE1_o => regOE1_s(2),
              reg2OE1_o => regOE1_s(3),
              reg3OE1_o => regOE1_s(4),

              reg0OE2_o => regOE2_s(1),
              reg1OE2_o => regOE2_s(2),
              reg2OE2_o => regOE2_s(3),
              reg3OE2_o => regOE2_s(4),

              accOE_o => accOE_s,
```

```
                    dinOE_o => dinOE_s,
                    irRST_o => irRST_s,
                    addrOE_o => addrOE_s,
                    readOE_o => read_o,
                    writeOE_o => writeOE_s) ;

    IR : ref16_c
        port map (pwr => pwr,
                  gnd => gnd,
                  d_i => ir_i,
                  rst_i => irRST_s,
                  clk_i => clk_i,
                  CE_i => irCE_s,
                  OE_i => pwr,
                  d_o => ir_s) ;

    AEB : buf12_c
        port map (pwr => pwr,
                  gnd => gnd,
                  d_i => ir_s (5 to 16),
                  OE_i => addrOE_s,
                  d_o => addr_o) ;

    write_o <=  writeOE_s after 4ns ;

--| when clk_i'changed then
--|     assert state.st = pla.state ;
--|     assert state.reg(0) = reg(0).state ;
--|     assert state.reg(1) = reg(1).state ;
--|     assert state.reg(2) = reg(2).state ;
--|     assert state.reg(3) = reg(3).state ;
--|     assert state.acc = acc.state ;
--|     assert state.ir = ir.state ;
--| end when ;

end structure ;
```

B.7 CPU configuration

```
use work.all ;

configuration cpuTC of cpu_tb is

    for structure
        for C : cpu_c
            use entity cpu(structure) ;
            for structure
                for l
                    for r : reg16_c
                        use entity reg16(structure) ;
                        for structure
                            for all : reg1_c
                                use entity reg1(structure) ;
                            end for ;
                        end for ;
                    end for ;
                end for ;
                for acc, ir : ref16_c
                    use entity ref16(structure) ;
                    for structure
                        for l
                            for all : ref1_c
                                use entity ref1(structure) ;
                            end for ;
                        end for ;
                    end for ;
                end for ;
                for C : pla_c
                    use entity pla(structure) ;
                    for structure
                        for all : ref1_c
                            use entity ref1(structure) ;
                        end for ;
                    end for ;
                end for ;
                for A : alu16_c
                    use entity alu16(structure) ;
                    for structure
                        for l
                            for all : alu1_c
                                use entity alu1(structure) ;
                            end for ;
                        end for ;
                    end for ;
                end for ;
```

```
                    for DOB, DIB : buf16_c
                        use entity buf16(structure) ;
                        for structure
                            for l
                                for all : buf1_c
                                    use entity buf1(structure) ;
                                end for ;
                            end for ;
                        end for ;
                    end for ;
                    for AEB : buf12_c
                        use entity buf12(structure) ;
                        for structure
                            for l
                                for all : buf1_c
                                    use entity buf1(structure) ;
                                end for ;
                            end for ;
                        end for ;
                    end for ;
                    for O0 : or16_c
                        use entity or16(structure) ;
                    end for ;
                    for O1, O2 : orr16_c
                        use entity orr16(structure) ;
                    end for ;
                end for ;
            end for ;
        end for ;

    end cpuTC ;
```

B.8 CPU support package

```
package cpu_p is

    type states is (if1, if2, if3, if4, ex1, ex2, ld, st) ;

    function intg ( inp : bit_vector )
        return integer ;
```

```
    function bitF ( x : integer ;
                    i : integer )
        return bit ;

    type int_at is array (0 to 3) of integer range 0 to 65535 ;

    type state_model is record
        st : states ;
        reg : int_at ;
        acc, ir : integer ;
    end record ;

    type reg16_t is bit_vector (1 to 16) ;
    type reg4_t is bit_vector (1 to 4) ;
    type reg2_t is bit_vector (1 to 2;

end cpu_p ;

package body cpu_p is

    function intg ( inp : bit_vector )
        return integer is

        variable temp : integer := 0 ;

    begin
        for i in inp'range loop
            if inp (i) = '1' then
                temp := 2*temp + 1 ;
            else
                temp := 2*temp ;
            end if ;
        end loop ;
    end intg ;

    function bitF ( x : integer ;
                    i : integer ) return bit is
        -- range assumed : 1 to 16
        variable y : integer := x ;
    begin
        for j in 15 downto i loop
            y := y / 2 ;
        end loop ;
        if y mod 2 = 0 then
            return '0' ;
```

```
        else
            return '1' ;
        end if ;
    end ;

end cpu_p ;
```

B.9 CPU test bench

```
use work.cpu_p.all ;

entity cpu_tb is

end cpu_tb ;

architecture structure of cpu_tb is

    signal tb_clk, tb_rst, tb_run, tb_clr : bit := '0' ;
    signal tb_din, tb_irin, tb_dout: bit_vector (1 to 16) := X"0000" ;
    signal tb_addr : bit_vector (1 to 12) := X"000" ;
    signal tb_read, tb_write : bit := '0' ;

    component cpu_c
      port ( pwr, gnd : in bit := '0' ;
             clk_i, rst_i : in bit := '0' ;
             run_i, clr_i : in bit := '0' ;
             d_i : in bit_vector (1 to 16) := X"0000" ;
             ir_i : in bit_vector (1 to 16) := X"0000" ;

             d_o : out bit_vector (1 to 16) := X"0000" ;
             addr_o : out bit_vector (1 to 12) := X"000" ;
             read_o, write_o : out bit := '0' ) ;
    end component ;

    signal run : boolean := true ;
    signal pwr, gnd : bit := '0' ;

begin
```

```
C : cpu_c
      port map ( clk_i => tb_clk,
                 clr_i => tb_clr,
                 rst_i => tb_rst,
                 run_i => tb_run,
                 d_i => tb_din,
                 ir_i => tb_irin,
                 d_o => tb_dout,
                 addr_o => tb_addr,
                 read_o => tb_read,
                 write_o => tb_write) ;

tb_clk <= TRANSPORT not tb_clk after 50ns ;

tb_clr <= TRANSPORT '0', '1' after 300ns,
                         '0' after 400ns ;
tb_run <= TRANSPORT '0', '1' after 500ns ;
tb_rst <= TRANSPORT '0', '1' after 100ns,
                         '0' after 200ns ;
tb_irin <= TRANSPORT
          B"0000_0000_0000_0000",
          B"1001_0011_0011_0011" after 600ns,
          -- ld : reg1 <= din
          B"1010_1100_1100_1100" after 1000ns,
          -- ld : reg2 <= din
          B"0101_1001_0000_0000" after 1400ns,
          -- ex : reg2 <= reg1 xor reg2
          B"1110_0011_0011_0011" after 1900ns ;
          -- st : dout <= reg2 ;
tb_din <= TRANSPORT
          X"0000",
          X"AFA0" after 600ns,
          X"55F5" after 1000ns ;

run <= false after 3000ns ;
assert run report "that's it folks" ;

pwr <= '1' ;

end structure ;
```

B.10 Or arrays

```
entity or16 is

    port ( pwr, gnd : in bit := '0' ;
           d1_i : in bit_vector (1 to 16) := X"0000" ;
           d2_i : in bit_vector (1 to 16) := X"0000" ;
           d_o : out bit_vector (1 to 16) := X"0000" ) ;

end or16 ;

architecture structure of or16 is

begin

    d_o (1) <= d1_i (1) or d2_i (1) after 1ns ;
    d_o (2) <= d1_i (2) or d2_i (2) after 1ns ;
    d_o (3) <= d1_i (3) or d2_i (3) after 1ns ;
    d_o (4) <= d1_i (4) or d2_i (4) after 1ns ;
    d_o (5) <= d1_i (5) or d2_i (5) after 1ns ;
    d_o (6) <= d1_i (6) or d2_i (6) after 1ns ;
    d_o (7) <= d1_i (7) or d2_i (7) after 1ns ;
    d_o (8) <= d1_i (8) or d2_i (8) after 1ns ;
    d_o (9) <= d1_i (9) or d2_i (9) after 1ns ;
    d_o (10) <= d1_i (10) or d2_i (10) after 1ns ;
    d_o (11) <= d1_i (11) or d2_i (11) after 1ns ;
    d_o (12) <= d1_i (12) or d2_i (12) after 1ns ;
    d_o (13) <= d1_i (13) or d2_i (13) after 1ns ;
    d_o (14) <= d1_i (14) or d2_i (14) after 1ns ;
    d_o (15) <= d1_i (15) or d2_i (15) after 1ns ;
    d_o (16) <= d1_i (16) or d2_i (16) after 1ns ;

end structure ;
```

```
entity orr16 is

    port ( pwr, gnd : in bit := '0' ;
           d1_i : in bit_vector (1 to 16) := B"00" ;
           d2_i : in bit_vector (1 to 16) := B"00" ;
```

```
            d3_i : in bit_vector (1 to 16) := B"00" ;
            d4_i : in bit_vector (1 to 16) := B"00" ;
            d_o : out bit_vector (1 to 16) := B"00" ) ;

end orr16 ;

architecture structure of orr16 is


begin

    d_o (1) <= d1_i (1) or d2_i (1) or d3_i (1) or d4_i (1) after 1ns ;
    d_o (2) <= d1_i (2) or d2_i (2) or d3_i (2) or d4_i (2) after 1ns ;
    d_o (3) <= d1_i (3) or d2_i (3) or d3_i (3) or d4_i (3) after 1ns ;
    d_o (4) <= d1_i (4) or d2_i (4) or d3_i (4) or d4_i (4) after 1ns ;
    d_o (5) <= d1_i (5) or d2_i (5) or d3_i (5) or d4_i (5) after 1ns ;
    d_o (6) <= d1_i (6) or d2_i (6) or d3_i (6) or d4_i (6) after 1ns ;
    d_o (7) <= d1_i (7) or d2_i (7) or d3_i (7) or d4_i (7) after 1ns ;
    d_o (8) <= d1_i (8) or d2_i (8) or d3_i (8) or d4_i (8) after 1ns ;
    d_o (9) <= d1_i (9) or d2_i (9) or d3_i (9) or d4_i (9) after 1ns ;
    d_o (10) <= d1_i (10) or d2_i (10) or d3_i (10) or d4_i (10) after 1ns ;
    d_o (11) <= d1_i (11) or d2_i (11) or d3_i (11) or d4_i (11) after 1ns ;
    d_o (12) <= d1_i (12) or d2_i (12) or d3_i (12) or d4_i (12) after 1ns ;
    d_o (13) <= d1_i (13) or d2_i (13) or d3_i (13) or d4_i (13) after 1ns ;
    d_o (14) <= d1_i (14) or d2_i (14) or d3_i (14) or d4_i (14) after 1ns ;
    d_o (15) <= d1_i (15) or d2_i (15) or d3_i (15) or d4_i (15) after 1ns ;
    d_o (16) <= d1_i (16) or d2_i (16) or d3_i (16) or d4_i (16) after 1ns ;


end structure ;
```

B.11 PLA

```
use work.cpu_p.all ;

entity pla is

    port ( pwr, gnd : in bit := '0' ;
           op_i : in bit_vector (1 to 2) ;
```

```
            r1_i : in bit_vector (1 to 2) ;
            r2_i :  in bit_vector (1 to 2) ;
            clr_i : in bit ;
            run_i : in bit ;
            rst_i : in bit ;
            clk_i : in bit ;
            irCE_o : out bit ;
            accCE_o : out bit ;
            reg0CE_o : out bit ;
            reg1CE_o : out bit ;
            reg2CE_o : out bit ;
            reg3CE_o : out bit ;

            reg0OE1_o  : out bit ;
            reg1OE1_o  : out bit ;
            reg2OE1_o  : out bit ;
            reg3OE1_o  : out bit ;

            reg0OE2_o  : out bit ;
            reg1OE2_o  : out bit ;
            reg2OE2_o  : out bit ;
            reg3OE2_o  : out bit ;

            accOE_o  : out bit ;
            dinOE_o  : out bit ;
            irRST_o  : out bit ;
            addrOE_o  : out bit ;
            readOE_o  : out bit ;
            writeOE_o  : out bit ) ;

     -- annotation

--| state model is states := if1 ;

begin
      -- annotation state update

--|     when rst_i = '1' then
--|         state <- if1 ;
--|     else
--|         when clk_i'changed ('1') then
--|             select state is
--|                 in if1 =>
--|                     when clr_i = '1' then
--|                         state <- if2 ;
```

```
--|                           elsewhen run_i = '1' then
--|                              state <- if3 ;
--|                           end when ;
--|                        in if2 =>
--|                           state <- if4 ;
--|                        in if3 =>
--|                           state <- if4 ;
--|                        in if4 =>
--|                           select op_i is
--|                              in X"0" => state <- if1 ;
--|                              in X"1" => state <- ex1 ;
--|                              in X"2" => state <- ld ;
--|                              in X"3" => state <- st ;
--|                           end select ;
--|                        in ex1 =>
--|                           state <- ex2 ;
--|                        in ex2 | ld | st =>
--|                           state <- if1 ;
--|                    end select ;
--|                 end when ;
--|              end when ;

            -- assertions
--|  when clk_i'changed ('1') then
--|      when (state = ex1 and r2_i = B"00") or
--|           (state = ld  and r1_i = B"00") then
--|          assert reg0CE_o = '1'
--|          report "pla_e : reg0CE_o /= '1' " ;
--|      else
--|          assert reg0CE_o = '0'
--|          report "pla_e : reg0CE_o /= '0' " ;
--|      end when ;
--|      when (state = ex1 and r2_i = B"01") or
--|           (state = ld  and r1_i = B"01") then
--|          assert reg1CE_o = '1'
--|          report "pla_e : reg1CE_o /= '1' " ;
--|      else
--|          assert reg1CE_o = '0'
--|          report "pla_e : reg1CE_o /= '0' " ;
--|      end when ;
--|      when (state = ex1 and r2_i = B"10") or
--|           (state = ld  and r1_i = B"10") then
--|          assert reg2CE_o = '1'
--|          report "pla_e : reg2CE_o /= '1' " ;
--|      else
--|          assert reg2CE_o = '0'
```

```
--|           report "pla_e : reg2CE_o /= '0' " ;
--|       end when ;
--|       when (state = ex1 and r2_i = B"11") or
--|            (state = ld  and r1_i = B"11") then
--|           assert reg3CE_o = '1'
--|           report "pla_e : reg3CE_o /= '1' " ;
--|       else
--|           assert reg3CE_o = '0'
--|           report "pla_e : reg3CE_o /= '0' " ;
--|       end when ;

--|       when (state = st) or (state = ex1) then
--|           select r1_i is
--|               in X"0" => assert reg0OE1_o = '1'
--|                          report "pla_e : reg0OE1_o /= '1' " ;
--|               in X"1" => assert reg1OE1_o = '1'
--|                          report "pla_e : reg1OE1_o /= '1' " ;
--|               in X"2" => assert reg2OE1_o = '1'
--|                          report "pla_e : reg2OE1_o /= '1' " ;
--|               in X"3" => assert reg3OE1_o = '1'
--|                          report "pla_e : reg3OE1_o /= '1' " ;
--|           end select ;
--|       else
--|           assert reg0OE1_o = '0'
--|           report "pla_e : reg0OE1_o /= '0' " ;
--|           assert reg1OE1_o = '0'
--|           report "pla_e : reg1OE1_o /= '0' " ;
--|           assert reg2OE1_o = '0'
--|           report "pla_e : reg2OE1_o /= '0' " ;
--|           assert reg3OE1_o = '0'
--|           report "pla_e : reg3OE1_o /= '0' " ;
--|       end when ;

--|       when (state = ex1) then
--|           select r2_i is
--|               in X"0" => assert reg0OE2_o = '1'
--|                          report "pla_e : reg0OE2_o /= '1' " ;
--|               in X"1" => assert reg1OE2_o = '1'
--|                          report "pla_e : reg1OE2_o /= '1' " ;
--|               in X"2" => assert reg2OE2_o = '1'
--|                          report "pla_e : reg2OE2_o /= '1' " ;
--|               in X"3" => assert reg3OE2_o = '1'
--|                          report "pla_e : reg3OE2_o /= '1' " ;
--|           end select ;
```

```
--|     else
--|         assert reg0OE2_o = '0'
--|         report "pla_e : reg0OE2_o /= '0' " ;
--|         assert reg1OE2_o = '0'
--|         report "pla_e : reg1OE2_o /= '0' " ;
--|         assert reg2OE2_o = '0'
--|         report "pla_e : reg2OE2_o /= '0' " ;
--|         assert reg3OE2_o = '0'
--|         report "pla_e : reg3OE2_o /= '0' " ;
--|     end when ;

--|     when (state = if3) then
--|         assert irCE_o = '1'
--|         report "pla_e : irCE_o /= '1' " ;
--|     else
--|         assert irCE_o = '0'
--|         report "pla_e : irCE_o /= '0' " ;
--|     end when ;

--|     when (state = if1) then
--|         assert accCE_o = '1'
--|         report "pla_e : accCE_o /= '1' " ;
--|     else
--|         assert accCE_o = '0'
--|         report "pla_e : accCE_o /= '0' " ;
--|     end when ;

--|     when (state = if2) then
--|         assert irRST_o = '1'
--|         report "pla_e : irRST_o /= '1' " ;
--|     else
--|         assert irRST_o = '0'
--|         report "pla_e : irRST_o /= '0' " ;
--|     end when ;

--|     when (state = ld or state = st) then
--|         assert addrOE_o = '1'
--|         report "pla_e : addrOE_o /= '1' " ;
--|     else
--|         assert addrOE_o = '0'
--|         report "pla_e : addrOE_o /= '0' " ;
--|     end when ;

--|     when (state = ld) then
--|         assert readOE_o = '1'
```

```
--|             report "pla_e : readOE_o /= '1' " ;
--|         else
--|             assert readOE_o = '0'
--|             report "pla_e : readOE_o /= '0' " ;
--|         end when ;

--|         when (state = st) then
--|             assert writeOE_o = '1'
--|             report "pla_e : writeOE_o /= '1' " ;
--|         else
--|             assert writeOE_o = '0'
--|             report "pla_e : writeOE_o /= '0' " ;
--|         end when ;

--|     end when ;


end pla ;


architecture structure of pla is

component ref1_c
    port ( pwr, gnd : in bit := '0' ;
           d_i : in bit := '0' ;
           rst_i : in bit := '0' ;
           clk_i : in bit := '0' ;
           CE_i : in bit := '0' ;
           OE_i : in bit := '0' ;
           d_o : out bit := '0' ) ;
    --| state model is bit := '0' ;
end component ;

signal state_s, newstate_s : bit_vector (2 DOWNTO 0) := B"000" ;
signal notstate_s : bit_vector (2 DOWNTO 0) := B"111" ;
signal notop_s, notr1_s, notr2_s : bit_vector (1 to 2) := B"11" ;
signal notclr_s : bit := '1' ;
signal a11_s, a12_s, a13_s, a14_s, a15_s : bit := '1' ;
signal a21_s, a24_s, a31_s : bit := '1' ;

signal a51_s, a52_s, a61_s, a62_s : bit := '1' ;
signal a71_s, a72_s, a81_s, a82_s : bit := '1' ;

signal b51_s, b52_s, b61_s, b62_s : bit := '1' ;
```

```
    signal b71_s, b72_s, b81_s, b82_s : bit := '1' ;

    signal c51_s, c61_s, c71_s, c81_s : bit := '1' ;

    signal a91_s, a01_s, a03_s, a04_s, a05_s : bit := '1' ;

begin

    rega : refl_c
        port map (d_i => newstate_s (0),
                  rst_i => rst_i,
                  clk_i => clk_i,
                  CE_i => pwr,
                  OE_i => pwr,
                  d_o => state_s (0)) ;

    regb : refl_c
        port map (d_i => newstate_s (1),
                  rst_i => rst_i,
                  clk_i => clk_i,
                  CE_i => pwr,
                  OE_i => pwr,
                  d_o => state_s (1)) ;

    regc : refl_c
        port map (d_i => newstate_s (2),
                  rst_i => rst_i,
                  clk_i => clk_i,
                  CE_i => pwr,
                  OE_i => pwr,
                  d_o => state_s (2)) ;

    notstate_s <= not state_s after 2ns ;
    notop_s <= not op_i after 2ns ;
    notr1_s <= not r1_i after 2ns ;
    notr2_s <= not r2_i after 2ns ;
    notclr_s <= not clr_i after 2ns ;

    a11_s <= TRANSPORT
        not (clr_i and notstate_s (1)
                   and notstate_s (2)) after 3ns ;
    a12_s <= TRANSPORT
        not (state_s (0) and notstate_s (1)
```

```
                          and notstate_s (2)) after 3ns ;
a13_s <= TRANSPORT
    not (notstate_s (0) and state_s (1)
                          and notstate_s (2)) after 3ns ;
a14_s <= TRANSPORT
    not (op_i (1) and op_i (2) and
         state_s (0) and state_s (1)
                        and notstate_s (2)) after 3ns ;
a15_s <= TRANSPORT
    not (notstate_s (0) and notstate_s (1)
                          and state_s (2)) after 3ns ;
a21_s <= TRANSPORT
    not (notclr_s and run_i and
         state_s (0) and notstate_s (1)
                        and notstate_s (2)) after 3ns ;
a24_s <= TRANSPORT
    not (op_i (2) and state_s (0)
                  and state_s (1)
                  and notstate_s (2)) after 3ns ;
a31_s <= TRANSPORT
    not (op_i (1) and state_s (0)
                  and state_s (1)
                  and notstate_s (2)) after 3ns ;

newstate_s (0) <= TRANSPORT
    not (a11_s and a12_s and a13_s
               and a14_s and a15_s) after 3 ns ;
newstate_s (1) <= TRANSPORT
    not (a12_s and a13_s and a21_s and a31_s) after 3ns ;
newstate_s (2) <= TRANSPORT
    not (a15_s and a24_s and a31_s) after 3ns ;

irCE_o <= not (a12_s and a13_s) after 3ns ;
irRST_o <= not a12_s after 2ns ;
accCE_o <= not a15_s after 2ns ;

a51_s <= TRANSPORT
    not (notr2_s (1) and notr2_s (2) and
         state_s (0) and notstate_s (1)
                     and state_s (2)) after 3ns ;
a52_s <= TRANSPORT
    not (notr1_s (1) and notr1_s (2) and
         notstate_s (0) and state_s (1)
                        and state_s (2)) after 3ns ;
```

```
a61_s <= TRANSPORT
    not (notr2_s (1) and r2_i (2) and
         state_s (0) and notstate_s (1)
                     and state_s (2)) after 3ns ;
a62_s <= TRANSPORT
    not (notr1_s (1) and r1_i (2) and
         notstate_s (0) and state_s (1)
                        and state_s (2)) after 3ns ;
a71_s <= TRANSPORT
    not (r2_i (1) and notr2_s (2) and
         state_s (0) and notstate_s (1)
                     and state_s (2)) after 3ns ;
a72_s <= TRANSPORT
    not (r1_i (1) and notr1_s (2) and
         notstate_s (0) and state_s (1)
                        and state_s (2)) after 3ns ;
a81_s <= TRANSPORT
    not (r2_i (1) and r2_i (2) and
         state_s (0) and notstate_s (1)
                     and state_s (2)) after 3ns ;
a82_s <= TRANSPORT
    not (r1_i (1) and r1_i (2) and
         notstate_s (0) and
         state_s (1) and state_s (2)) after 3ns ;

reg0CE_o <= not (a51_s and a52_s) after 3ns ;
reg1CE_o <= not (a61_s and a62_s) after 3ns ;
reg2CE_o <= not (a71_s and a72_s) after 3ns ;
reg3CE_o <= not (a81_s and a82_s) after 3ns ;

b51_s <= not (state_s (2) and notstate_s (1) and
              notstate_s (0) and
              notr1_s (1) and notr1_s (2)) ;
b52_s <= not (state_s (2) and state_s (1) and
              state_s (0) and
              notr1_s (1) and notr1_s (2)) ;
b61_s <= not (state_s (2) and notstate_s (1) and
              notstate_s (0) and
              r1_i (1) and notr1_s (2)) ;
b62_s <= not (state_s (2) and state_s (1) and
              state_s (0) and
              r1_i (1) and notr1_s (2)) ;
b71_s <= not (state_s (2) and notstate_s (1) and
```

```
                    notstate_s (0) and
                    notr1_s (1) and r1_i (2)) ;
b72_s <= not (state_s (2) and state_s (1) and
                    state_s (0) and
                    notr1_s (1) and r1_i (2)) ;
b81_s <= not (state_s (2) and notstate_s (1) and
                    notstate_s (0) and
                    r1_i (1) and r1_i (2)) ;
b82_s <= not (state_s (2) and state_s (1) and
                    state_s (0) and
                    r1_i (1) and r1_i (2)) ;

reg0OE1_o <= not (b51_s and b52_s) after 3ns ;
reg1OE1_o <= not (b61_s and b62_s) after 3ns ;
reg2OE1_o <= not (b71_s and b72_s) after 3ns ;
reg3OE1_o <= not (b81_s and b82_s) after 3ns ;

c51_s <= not (state_s (2) and notstate_s (1)
                              and notstate_s (0) and
                    notr2_s (1) and notr2_s (2)) ;
c61_s <= not (state_s (2) and notstate_s (1)
                              and notstate_s (0) and
                    r2_i (1) and notr2_s (2)) ;
c71_s <= not (state_s (2) and notstate_s (1)
                              and notstate_s (0) and
                    notr2_s (1) and r2_i (2)) ;
c81_s <= not (state_s (2) and notstate_s (1)
                              and notstate_s (0) and
                    r2_i (1) and r2_i (2)) ;


reg0OE2_o <= not (c51_s) after 3ns ;
reg1OE2_o <= not (c61_s) after 3ns ;
reg2OE2_o <= not (c71_s) after 3ns ;
reg3OE2_o <= not (c81_s) after 3ns ;


a91_s <= not (state_s (0) and notstate_s (1)
                              and state_s (2)) after 3ns ;
accOE_o <= not a91_s after 2ns ;
a01_s <= not (notstate_s (0) and state_s (1)
                                    and state_s (2)) after 3ns ;
dinOE_o <= not a01_s after 2ns ;
```

```
    a03_s <= not (notstate_s (0) and state_s (1)
                                    and state_s (2)) after 3ns ;
    readOE_o <= not a03_s after 2ns ;
    a04_s <= not (state_s (1) and state_s (2)) after 3ns ;
    addrOE_o <= not a04_s after 2ns ;
    a05_s <= not (state_s (0) and state_s (1)
                                and state_s (2)) after 3ns ;
    writeOE_o <= not a05_s after 2ns ;


    -- mapping

--| when clk_i'changed ('1') then
--|     select state_s is
--|         in B"000" => assert state = if1 ;
--|         in B"001" => assert state = if2 ;
--|         in B"010" => assert state = if3 ;
--|         in B"011" => assert state = if4 ;
--|         in B"100" => assert state = ex1 ;
--|         in B"101" => assert state = ex2 ;
--|         in B"110" => assert state = ld ;
--|         in B"111" => assert state = st ;
--|     end select ;
--| end when ;


end structure ;
```

B.12 One bit one output register

```
use work.cpu_p.all ;

entity ref1 is

    port ( pwr, gnd : in bit := '0' ;
           d_i : in bit := '0' ;
           rst_i : in bit := '0' ;
           clk_i : in bit := '0' ;
```

```
                CE_i : in bit := '0' ;
                OE_i : in bit := '0' ;
                d_o : out bit := '0' ) ;

        -- Annotation

    --| state model is bit := '0' ;

    begin

        -- state update

    --| when clk_i'changed('1') then
    --|     when CE_i = '1' then
    --|         when rst_i = '1' then
    --|             state <- '0' ;
    --|         else
    --|             state <- d_i ;
    --|         end when ;
    --|     end when ;
    --| end when ;

        -- assertions

    --| when clk_i'changed('1') then
    --|     when OE_i = '1' then
    --|         assert d_o = state
    --|         report "ref1 : d_o /= state" ;
    --|     end when ;
    --| end when ;

    end ref1 ;


    architecture structure of ref1 is

        signal i00_s, i01_s : bit := '1' ;

        signal n011_s, n012_s : bit := '1' ;
        signal n021_s : bit := '0' ;

        signal n11_s, n21_s : bit := '0' ;
        signal n12_s, n13_s, n14_s, n22_s : bit := '1' ;

        signal n31_s : bit := '1' ;
```

```
    signal i41_s : bit := '0' ;

begin

    i00_s <= not CE_i after 2ns ;
    i01_s <= not rst_i after 2ns ;
    n011_s <= not (d_i and CE_i and i01_s) after 3ns ;
    n012_s <= not (i00_s and n21_s) after 3ns ;
    n021_s <= not (n012_s and n011_s) after 3ns ;

    n11_s <= not (n12_s and n14_s) after 3ns ;
    n12_s <= not (n11_s and clk_i) after 3ns ;
    n13_s <= not (n12_s and clk_i and n14_s) after 3ns ;
    n14_s <= not (n13_s and n021_s) after 3ns ;

    n21_s <= not (n12_s and n22_s) after 3ns ;
    n22_s <= not (n21_s and n13_s) after 3ns ;

    n31_s <= not (n21_s and OE_i) after 3ns ;
    d_o <= not n31_s after 2ns ;

end structure ;
```

B.13 16 bit one output register

```
use work.cpu_p.all ;

entity ref16 is

    port ( pwr, gnd : in bit := '0' ;
           d_i : in bit_vector (1 to 16) := X"0000" ;
           rst_i : in bit := '0' ;
           clk_i : in bit := '0' ;
           CE_i : in bit := '0' ;
           OE_i : in bit := '0' ;
           d_o : out bit_vector (1 to 16) := X"0000" ) ;

    -- Annotation
```

```
--| state model is integer range 0 to 65535 := 0 ;

begin

    -- state update

--| when clk_i'changed('1') then
--|     when CE_i = '1' then
--|         when rst_i = '1' then
--|             state <- 0 ;
--|         else
--|             state <- intg (d_i) ;
--|         end when ;
--|     end when ;
--| end when ;

    -- assertions

--| when clk_i'changed('1') then
--|     when OE_i = '1' then
--|         assert intg (d_o) = state
--|         report "not d_o = state" ;
--|     end when ;
--| end when ;

end ref16 ;


architecture structure of ref16 is

    component ref1_c
        port ( pwr, gnd : in bit := '0' ;
               d_i : in bit := '0' ;
               rst_i : in bit := '0' ;
               clk_i : in bit := '0' ;
               CE_i : in bit := '0' ;
               OE_i : in bit := '0' ;
               d_o : out bit := '0' ) ;
        --| state model is bit := '0' ;
    end component ;

begin

    l : for i in 1 to 16 generate
```

```
        R : refl_c
            port map (pwr => pwr,
                      gnd => gnd,
                      d_i => d_i(i),
                      rst_i => rst_i,
                      clk_i => clk_i,
                      CE_i => CE_i,
                      OE_i => OE_i,
                      d_o => d_o(i)) ;

    end generate ;


    -- mapping

--| when clk_i'changed('1') then
--|     for i in 1 to 16 generate
--|         assert bitF (state, i) = ref_i(i).state ;
--|     end generate ;
--| end when ;

end structure ;
```

B.14 One bit two output register

```
use work.cpu_p.all ;

entity reg1 is

    port ( pwr, gnd : in bit := '0' ;
           d_i : in bit := '0' ;
           rst_i : in bit := '0' ;
           clk_i : in bit := '0' ;
           CE_i : in bit := '0' ;
           OE1_i : in bit := '0' ;
           d1_o : out bit := '0' ;
           OE2_i : in bit := '0' ;
           d2_o : out bit := '0' ) ;
```

```
    -- Annotation

--| state model is bit := '0' ;

begin

    -- state update

--| when clk_i'changed('1') then
--|     when CE_i = '1' then
--|         when rst_i = '1' then
--|             state <- '0' ;
--|         else
--|             state <- d_i ;
--|         end when ;
--|     end when ;
--| end when ;

    -- assertions

--| when clk_i'changed('1') then
--|     when OE1_i = '1' then
--|         assert d1_o = state
--|         report "reg1 : d1_o /= state" ;
--|     else
--|         assert d1_o = '0'
--|         report "reg1 : d1_o /= 0" ;
--|     end when ;
--|     when OE2_i = '1' then
--|         assert d2_o = state
--|         report "reg1 : d2_o /= state" ;
--|     else
--|         assert d2_o = '0'
--|         report "reg1 : d2_o /= 0" ;
--|     end when ;
--| end when ;

end reg1 ;



architecture structure of reg1 is

    signal i00_s, i01_s : bit := '1' ;
```

```
    signal n011_s, n012_s : bit := '1' ;
    signal n021_s : bit := '0' ;

    signal n11_s, n21_s : bit := '0' ;
    signal n12_s, n13_s, n14_s, n22_s : bit := '1' ;

    signal n31_s : bit := '1' ;
    signal n32_s : bit := '1' ;
    signal i41_s : bit := '0' ;

begin

    i00_s <= not CE_i after 2ns ;
    i01_s <= not rst_i after 2ns ;
    n011_s <= not (d_i and CE_i and i01_s) after 3ns ;
    n012_s <= not (i00_s and n21_s) after 3ns ;
    n021_s <= not (n012_s and n011_s) after 3ns ;

    n11_s <= not (n12_s and n14_s) after 3ns ;
    n12_s <= not (n11_s and clk_i) after 3ns ;
    n13_s <= not (n12_s and clk_i and n14_s) after 3ns ;
    n14_s <= not (n13_s and n021_s) after 3ns ;

    n21_s <= not (n12_s and n22_s) after 3ns ;
    n22_s <= not (n21_s and n13_s) after 3ns ;

    n31_s <= not (n21_s and OE1_i) after 3ns ;
    d1_o <= not n31_s after 2ns ;
    n32_s <= not (n21_s and OE2_i) after 3ns ;
    d2_o <= not n32_s after 2ns ;

end structure ;
```

B.15 16 bit two output register

```
use work.cpu_p.all ;

entity reg16 is
```

```
    port ( pwr, gnd : in bit := '0' ;
           d_i : in bit_vector (1 to 16) := X"0000" ;
           rst_i : in bit := '0' ;
           clk_i : in bit := '0' ;
           CE_i : in bit := '0' ;
           OE1_i : in bit := '0' ;
           d1_o : out bit_vector (1 to 16) := X"0000" ;
           OE2_i : in bit := '0' ;
           d2_o : out bit_vector (1 to 16) := X"0000" ) ;

    -- Annotation

--| state model is integer range 0 to 65535 := 0 ;

begin

    -- state update

--| when clk_i'changed('1') then
--|     when CE_i = '1' then
--|         when rst_i = '1' then
--|             state <- 0 ;
--|         else
--|             state <- intg (d_i) ;
--|         end when ;
--|     end when ;
--| end when ;

    -- assertions

--| when clk_i'changed('1') then
--|     when OE1_i = '1' then
--|         assert intg (d1_o) = state
--|         report "not d1_o = state" ;
--|     else
--|         assert int (d1_o) = 0
--|         report "not d1_o = 0" ;
--|     end when ;
--|     when OE2_i = '1' then
--|         assert intg (d2_o) = 0 ;
--|         report "not d2_o = 0" ;
--|     else
--|     end when ;
--| end when ;
```

```
end reg16 ;

architecture structure of reg16 is

    component reg1_c
        port ( pwr, gnd : in bit := '0' ;
               d_i : in bit := '0' ;
               rst_i : in bit := '0' ;
               clk_i : in bit := '0' ;
               CE_i : in bit := '0' ;
               OE1_i : in bit := '0' ;
               d1_o : out bit := '0' ;
               OE2_i : in bit := '0' ;
               d2_o : out bit := '0' ) ;
        --| state model is bit := '0' ;
    end component ;

begin

    l : for i in 1 to 16 generate

        R : reg1_c
            port map (pwr => pwr,
                      gnd => gnd,
                      d_i => d_i(i),
                      rst_i => rst_i,
                      clk_i => clk_i,
                      CE_i => CE_i,
                      OE1_i => OE1_i,
                      d1_o => d1_o(i),
                      OE2_i => OE2_i,
                      d2_o => d2_o(i)) ;

    end generate ;


    -- mapping

--| when clk_i'changed('1') then
--|     for i in 1 to 16 generate
--|         assert bitF (state, i) = reg_i(i).state ;
--|     end generate ;
--| end when ;
```

```
end structure ;
```

Bibliography

[1] L. M. Augustin. An algebra of waveforms. In *Proceedings of the IFIP International Workshop on Applied Formal Methods For Correct VLSI Design*, L. Claesen, editor, pages 159-168, Leuven, Belgium, November 1989.

[2] L. M. Augustin. Timing models in VAL/VHDL. In *International Conference on Computer-Aided Design (ICCAD '89) Digest of Technical Papers*, pages 122-125, Santa Clara, CA, November 1989.

[3] L. M. Augustin, B. A. Gennart, Y. Huh, D. C. Luckham, and A. G. Stanculescu. VAL: An annotation language for VHDL. In *International Conference on Computer-Aided Design (ICCAD '87) Digest of Technical Papers*, pages 418-421, Santa Clara, CA, November 1987.

[4] L. M. Augustin, B. A. Gennart, Y. Huh, D. C. Luckham, and A. G. Stanculescu. Verification of VHDL designs using VAL. In *Proceedings of the 25th Design Automation Conference (DAC)*, pages 48-53, Anaheim, CA, June 1988.

[5] D. Coelho. *The VHDL Handbook.* Volume , Kluwer Academic Publishers, edition, 1989.

[6] O. J. Dahl. *Can Program Proving be Made Practical?* Technical Report 33, Institute of Informatics, University of Oslo, May 1978.

[7] J. A. Darringer. The application of program verification techniques to hardware verification. In *Proceedings of the 16th Design Automation Conference (DAC)*, pages 375-381, 1979.

[8] R. P. Gabriel. Draft report on requirements for a common prototyping system. *SIGPLAN Notices*, 24(3):93ff, March 1989.

[9] M. J. C. Gordon. *How to Specify and Verify Hardware Using Higher Order Logic.* Lecture Notes, University of Texas at Austin, Autumn 1984.

[10] J. V. Guttag. Abstract data types and the development of data structures. *Communications of the ACM*, 6(2):396-404, June 1977.

[11] D. Helmbold and D. Luckham. *Debugging Ada Tasking Programs.* Technical Report CSL-84-262, Computer Systems Laboratory, Stanford University, Stanford, CA, July 1984.

[12] C. A. R. Hoare. Hints on programming language design. In *Sigact/Sigplan Symposium on Principles of Programming Languages*, October 1973. Also published as Stanford University Computer Science Department Technical Report No. CS-73-403, Dec. 1973.

[13] Y. Huh. *Formal Specification and Verification of Hierarchical VLSI.* PhD thesis, Stanford University, Department of Electrical Engineering, Stanford, CA, December 1985.

[14] D. Ku and G. De Micheli. *Hardware C – A Language for Hardware Design.* Technical Report CSL-TR-88-362, Computer Systems Laboratory, Stanford University, Stanford, CA 94305, August 1988.

[15] R. Lipsett, C. Schaefer, and C. Ussery. *VHDL: Hardware Description and Design.* Volume , Kluwer Academic Publishers, edition, 1989.

[16] D. C. Luckham, Y. Huh, and A. G. Stanculescu. *The Semantics of Timing Constructs in Hardware Description Languages.* Technical Report CSL-TR-86-303, Computer Systems Laboratory, Stanford University, August 1986.

[17] D. C. Luckham, A. Stanculescu, Y. Huh, and S.Ghosh. The semantics of timing constructs in hardware description languages. In *IEEE International Conference on Computer Design: VLSI in Computers*

(ICCD '86), pages 10-14, Port Chester, New York, October 1986. Also published as Stanford Univerity Computer Systems Laboratory Technical Report CSL-TR-86-303.

[18] D. C. Luckham and F.W. vonHenke. An overview of ANNA, a specification language for Ada. *IEEE Software*, 2(2):9-22, March 1985.

[19] C. Mead and L. Conway. *Introduction to VLSI Systems.* Addison-Wesley, 1980.

[20] S. Sankar. *Automatic Runtime Consistency Checking and Debugging of Annotated Programs (An Overview).* Technical Report under preparation, Computer Systems Laboratory, Stanford University, Stanford, CA, 1988.

[21] S. Sankar. *A Note on the Detection of an Ada Compiler Bug while Debugging an Anna Program.* Technical Report under preparation, Computer Systems Laboratory, Stanford University, Stanford, CA, 1988.

[22] M. Shahdad, R. Lipsett, E. Marschner, K. Sheehan, H. Cohen, R. Waxman, and D. Ackley. VHSIC hardware description language. *Computer*, 18(2):94-103, February 1985.

[23] *VHDL Design Analysis and Justification.* Intermetrics, July 1984. IR-MD-018-1.

[24] *VHDL Language Reference Manual.* October 1986. IEEE Preliminary Version 7.2.

[25] *IEEE Standard VHDL Language Reference Manual.* IEEE, Inc., 345 East 47th Street, New York, NY, 10017, March 1987. IEEE Standard 1076-1987.

[26] *VHDL Language Refinement Rationale.* CAD Language Systems, Inc., Rockville, MD, March 1987.

[27] *The TTL Data Book.* Texas Instruments, Inc., P. O. Box 225012, Dallas, TX 75265, second edition, 1981.

Index